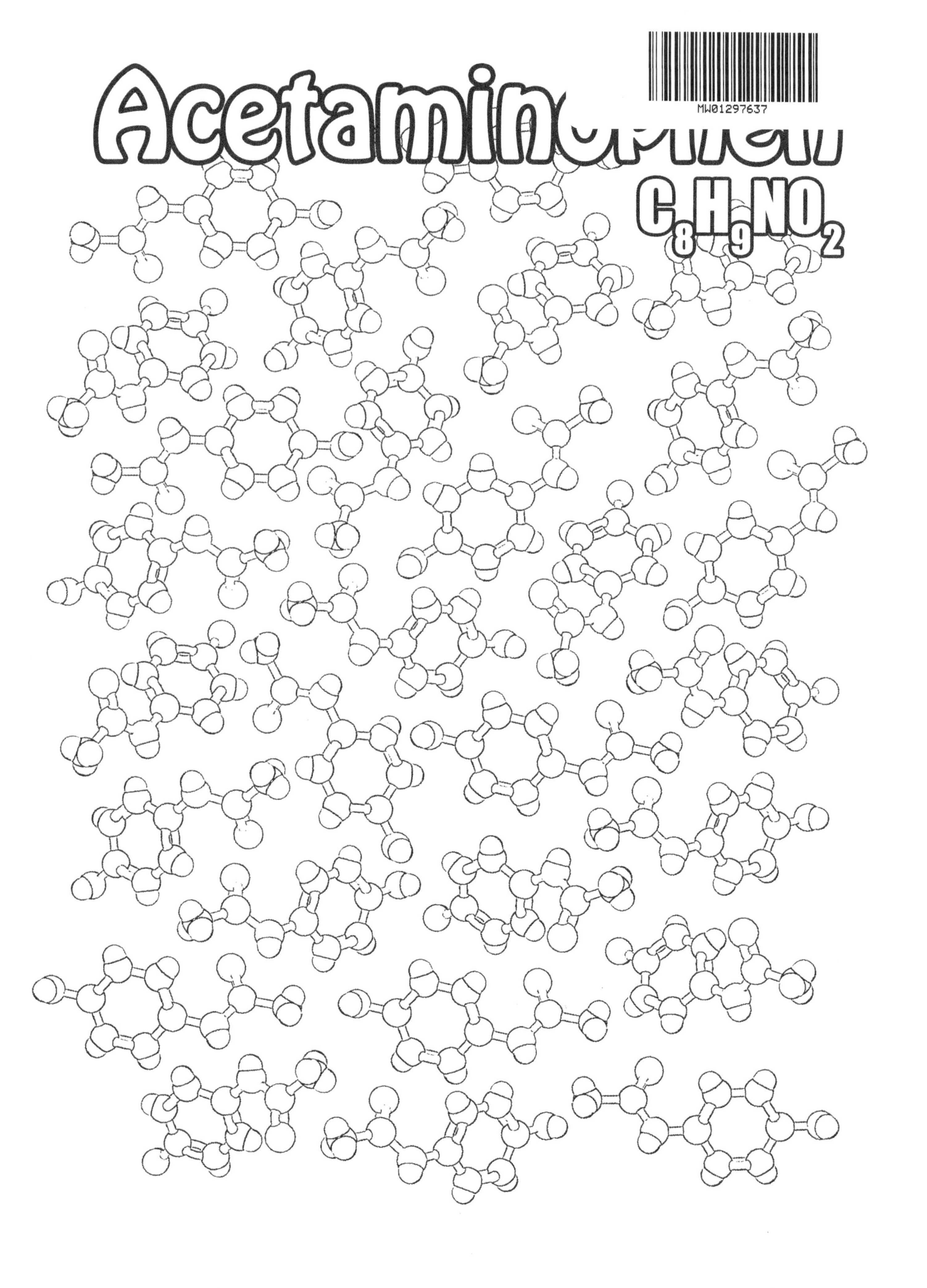
Acetaminophen
$C_8H_9NO_2$
MW01297637

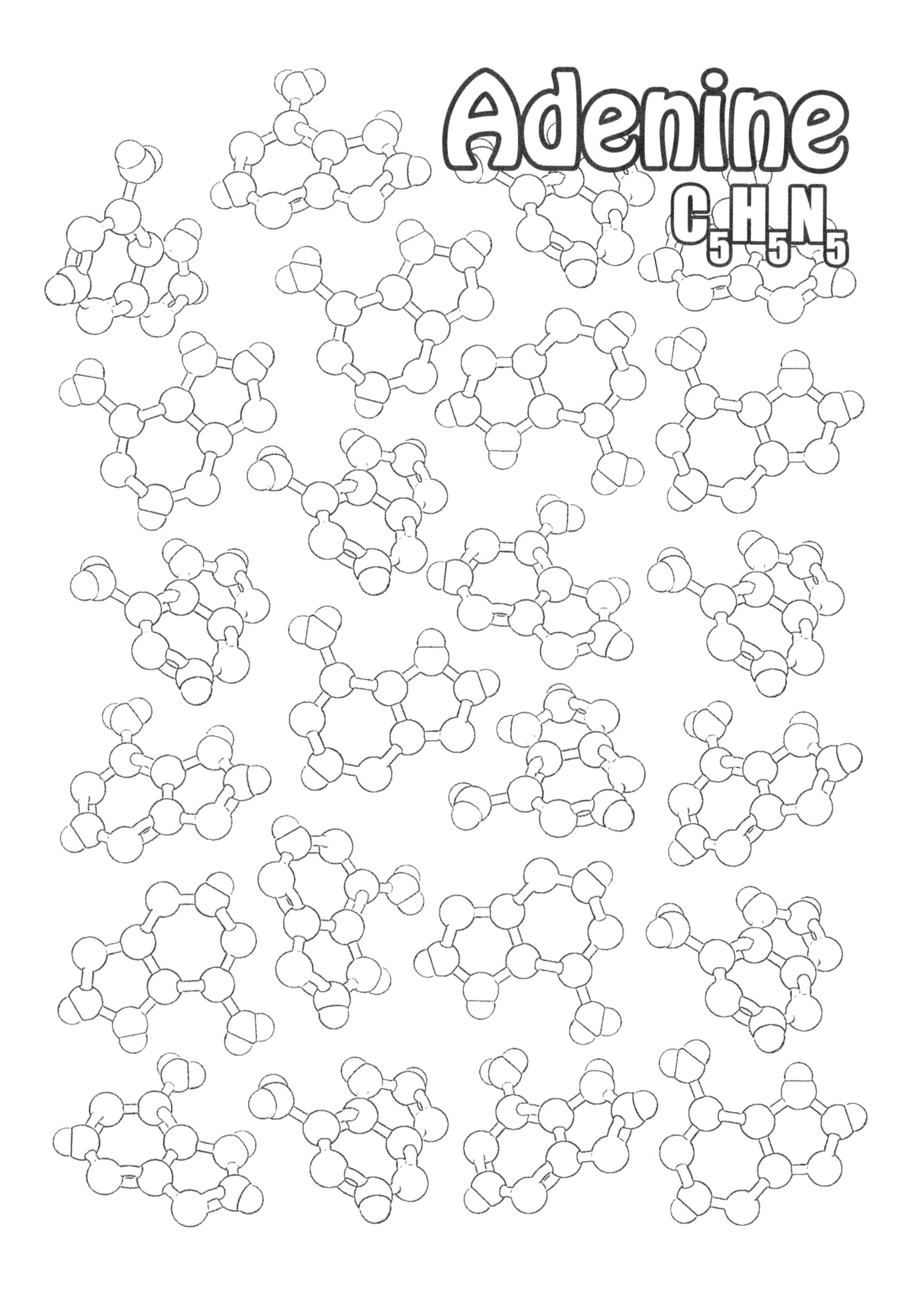
Adenine
$C_5H_5N_5$

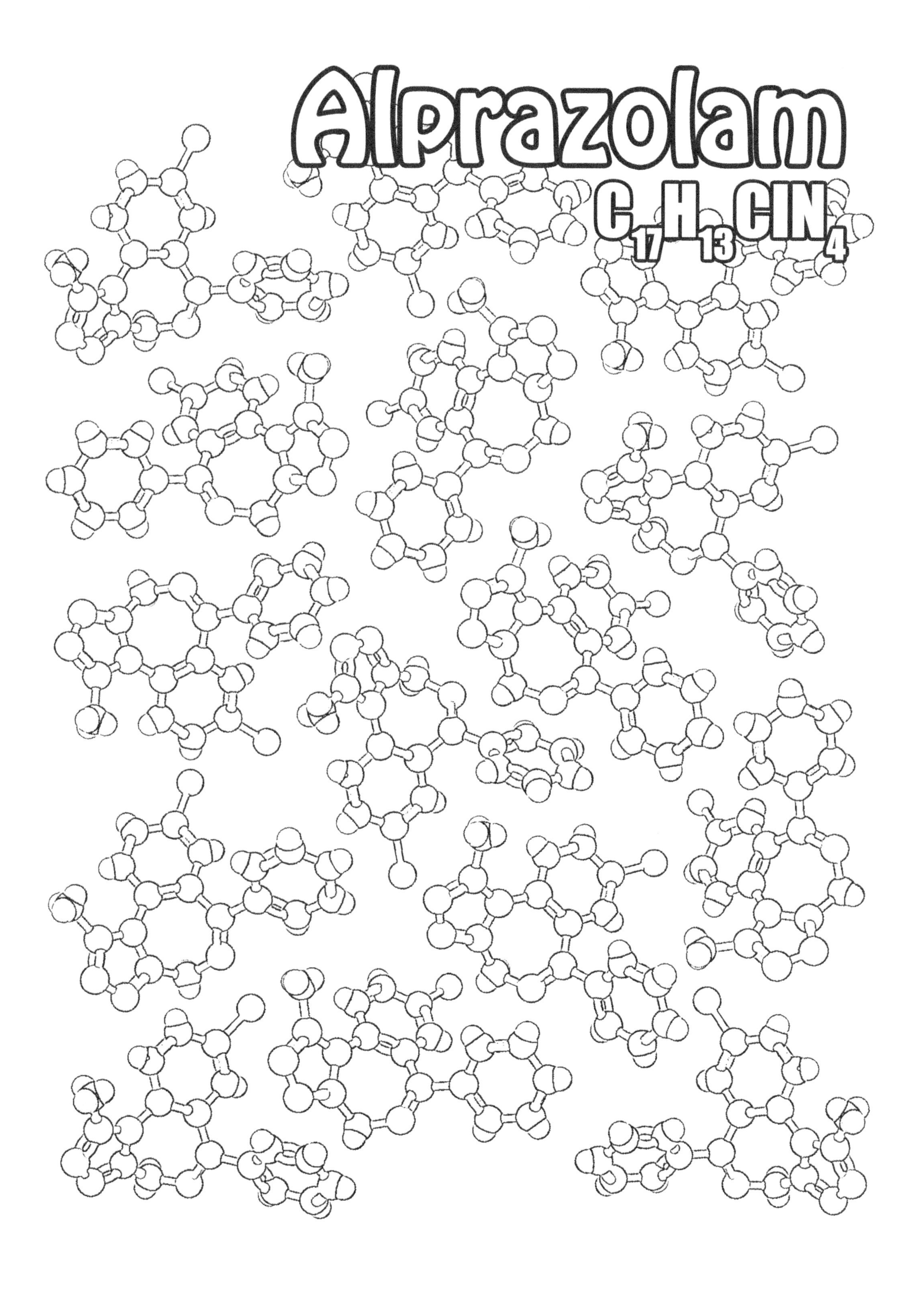
Alprazolam
C17H13ClN4

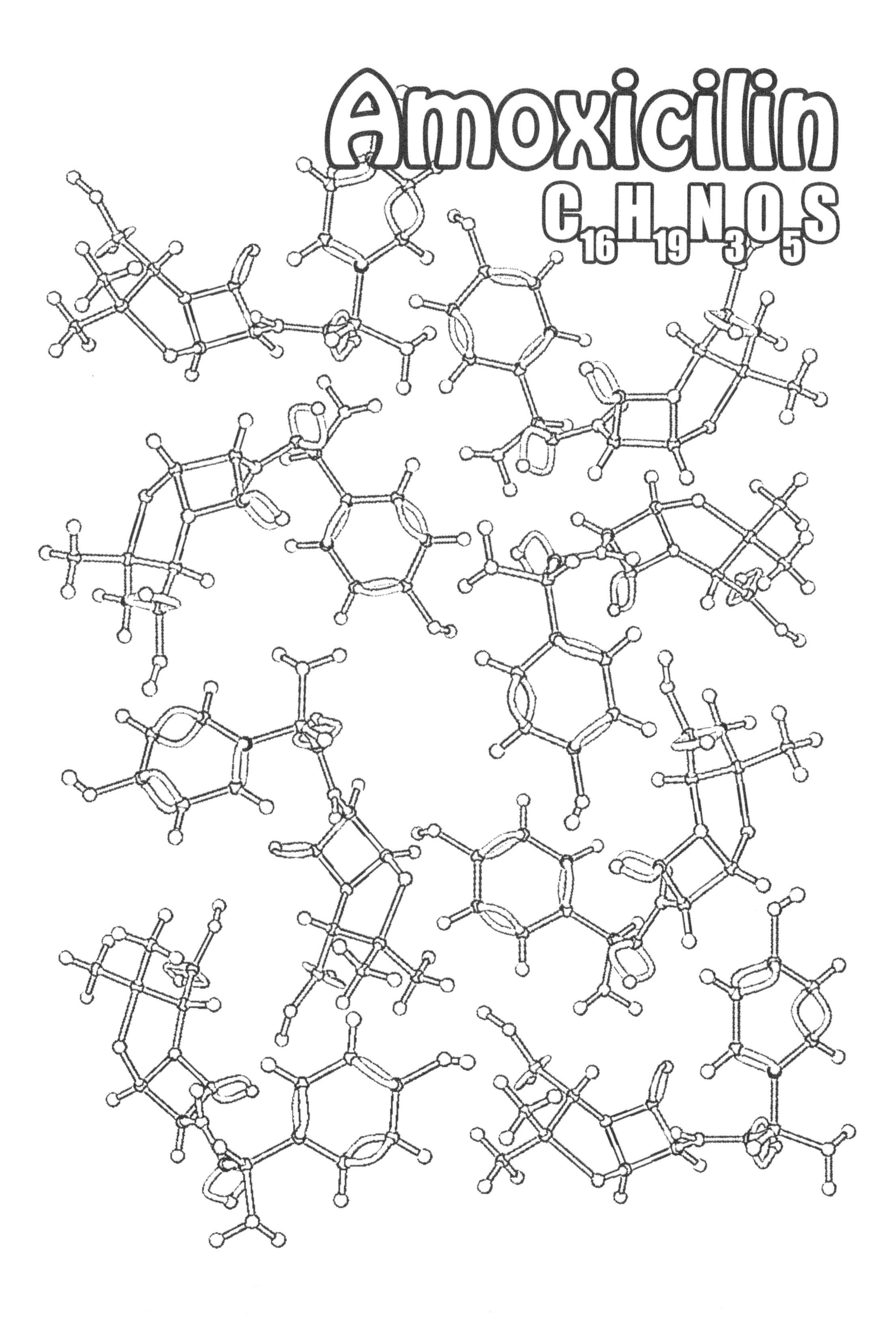
Amoxicilin
$C_{16}H_{19}N_3O_5S$

Bacitracin
$C_{66}H_{103}N_{17}O_{16}S$

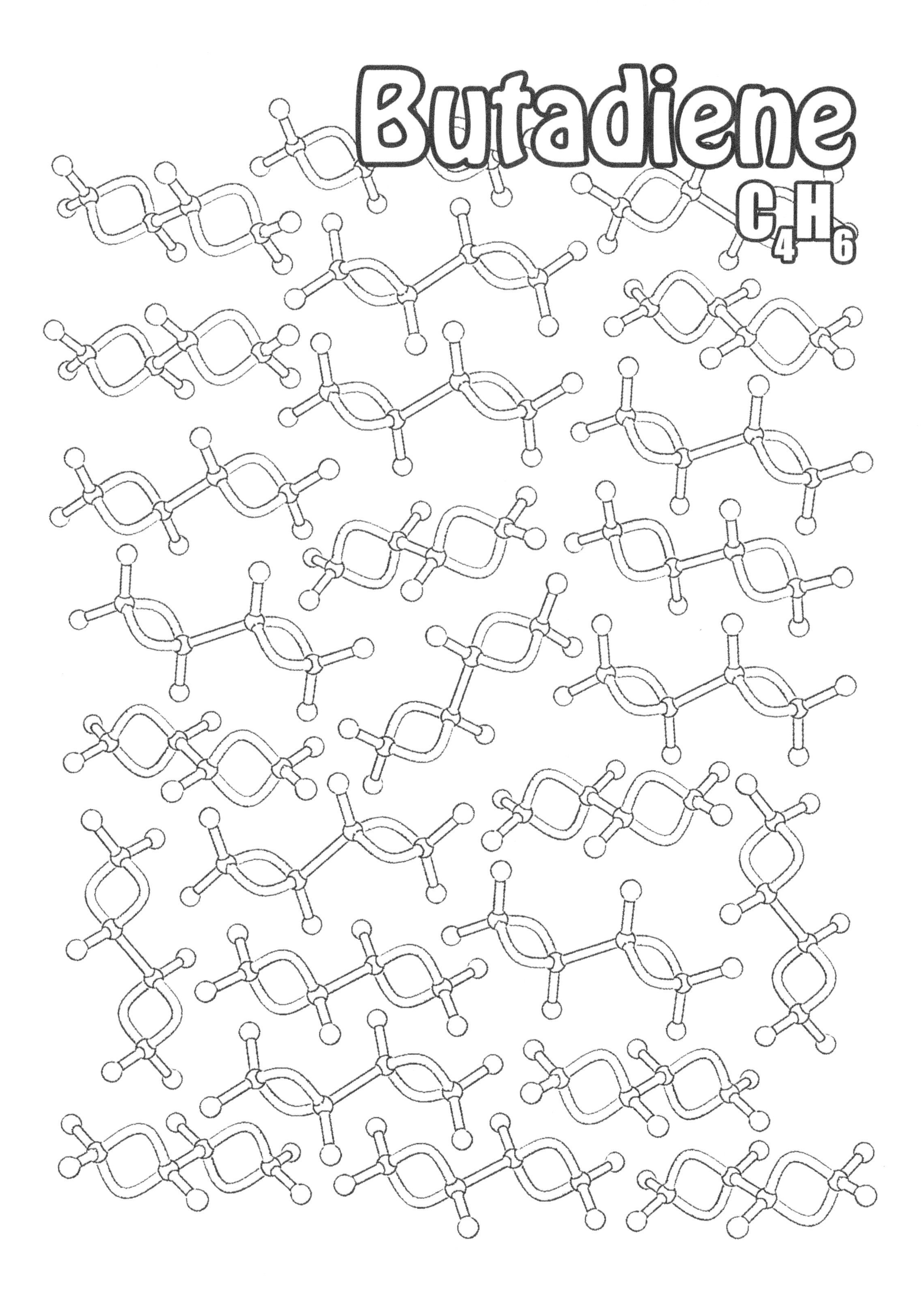
Butadiene
C_4H_6

Chloroform
$CHCl_3$

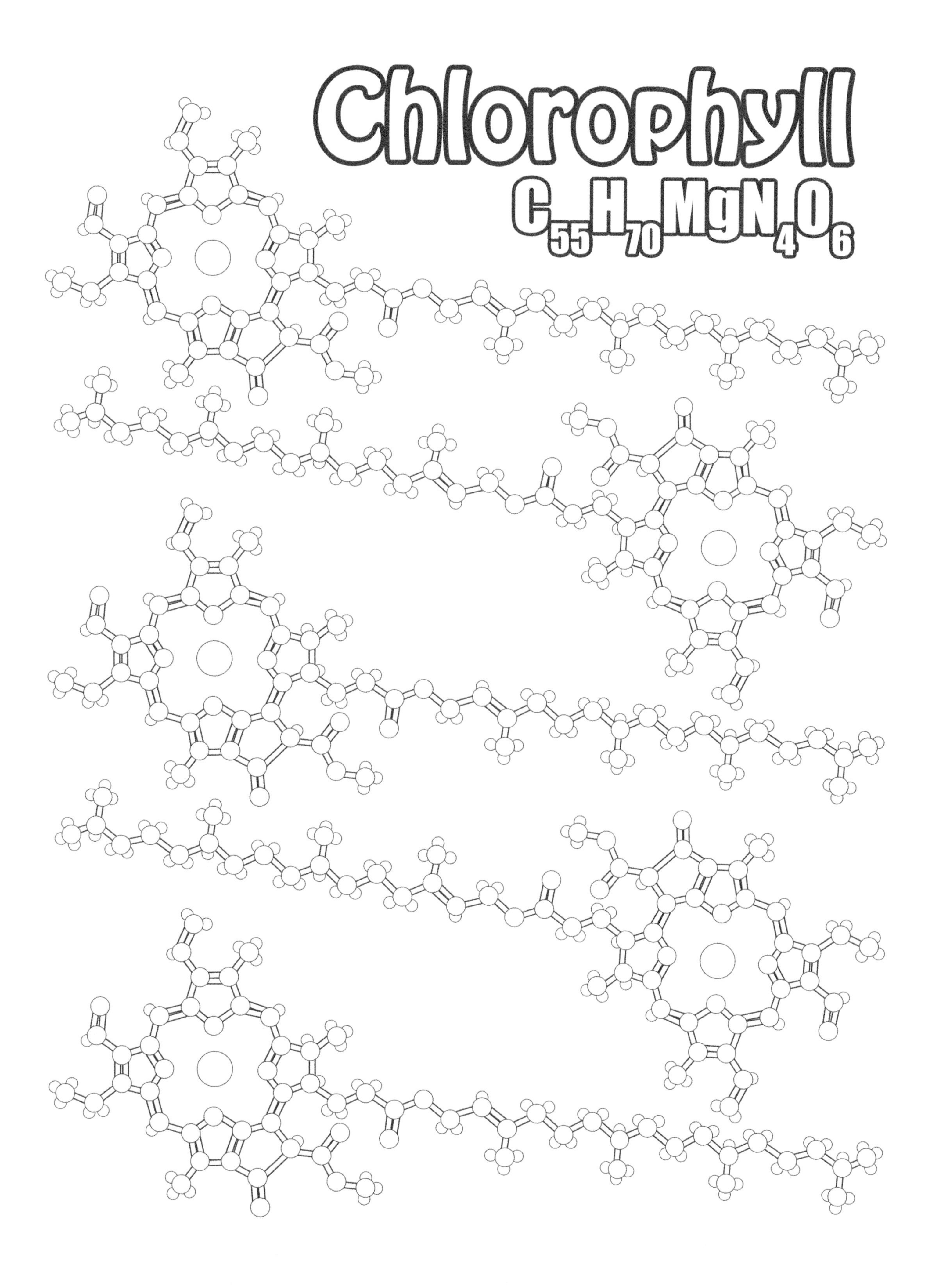
Chlorophyll
$C_{55}H_{70}MgN_4O_6$

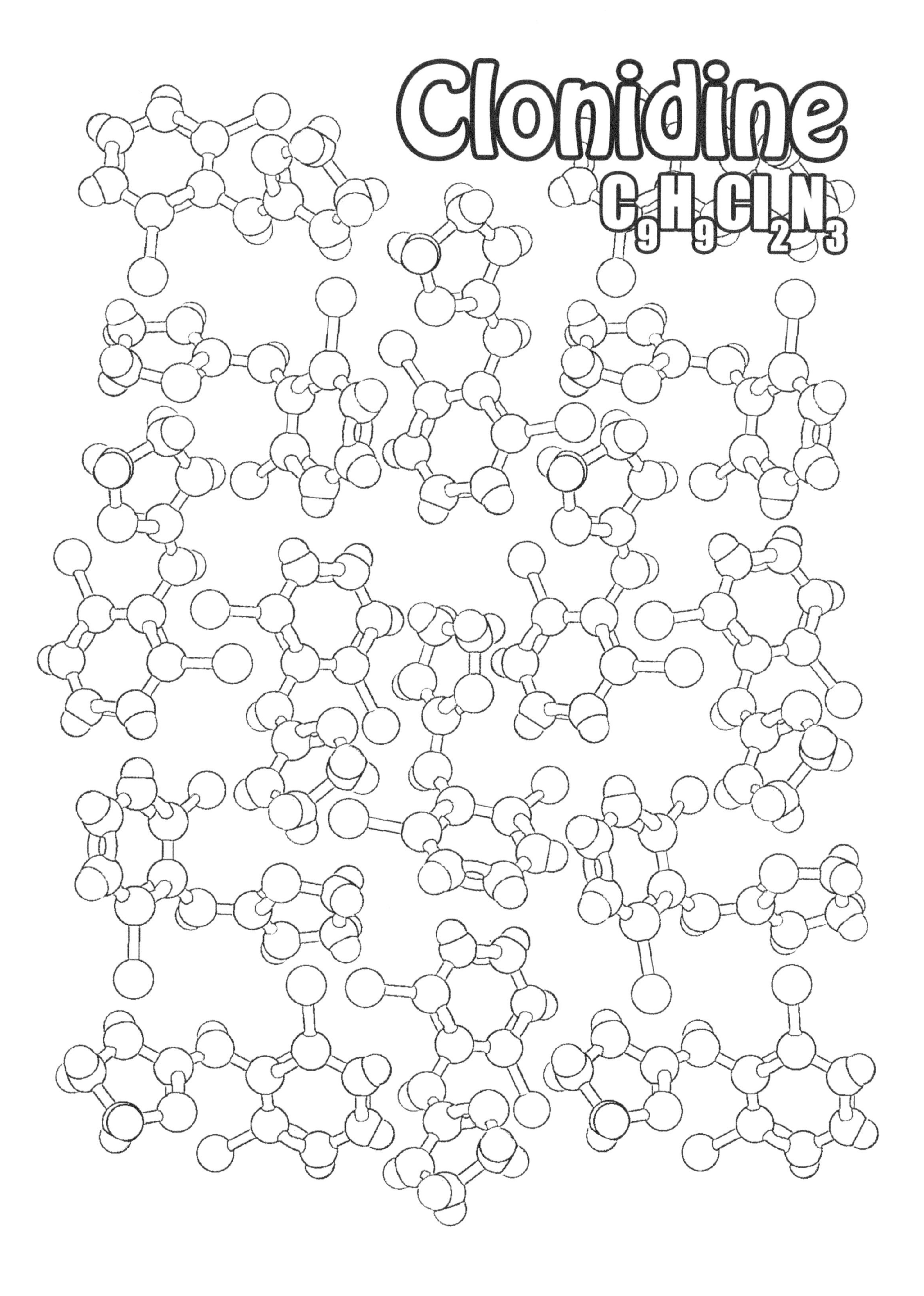
Clonidine
C9H9Cl2N3

Cytosine

$C_4H_5N_3O$

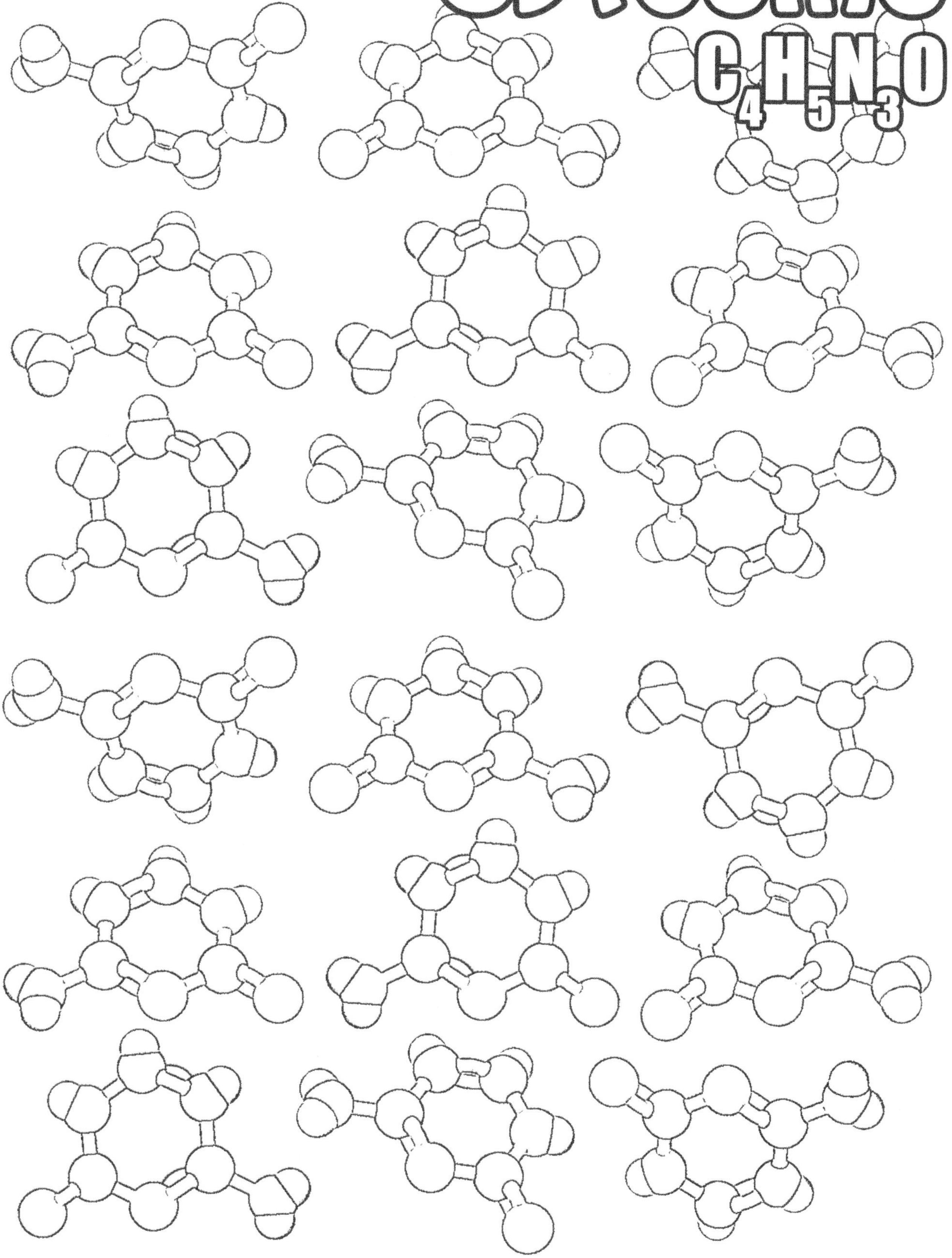

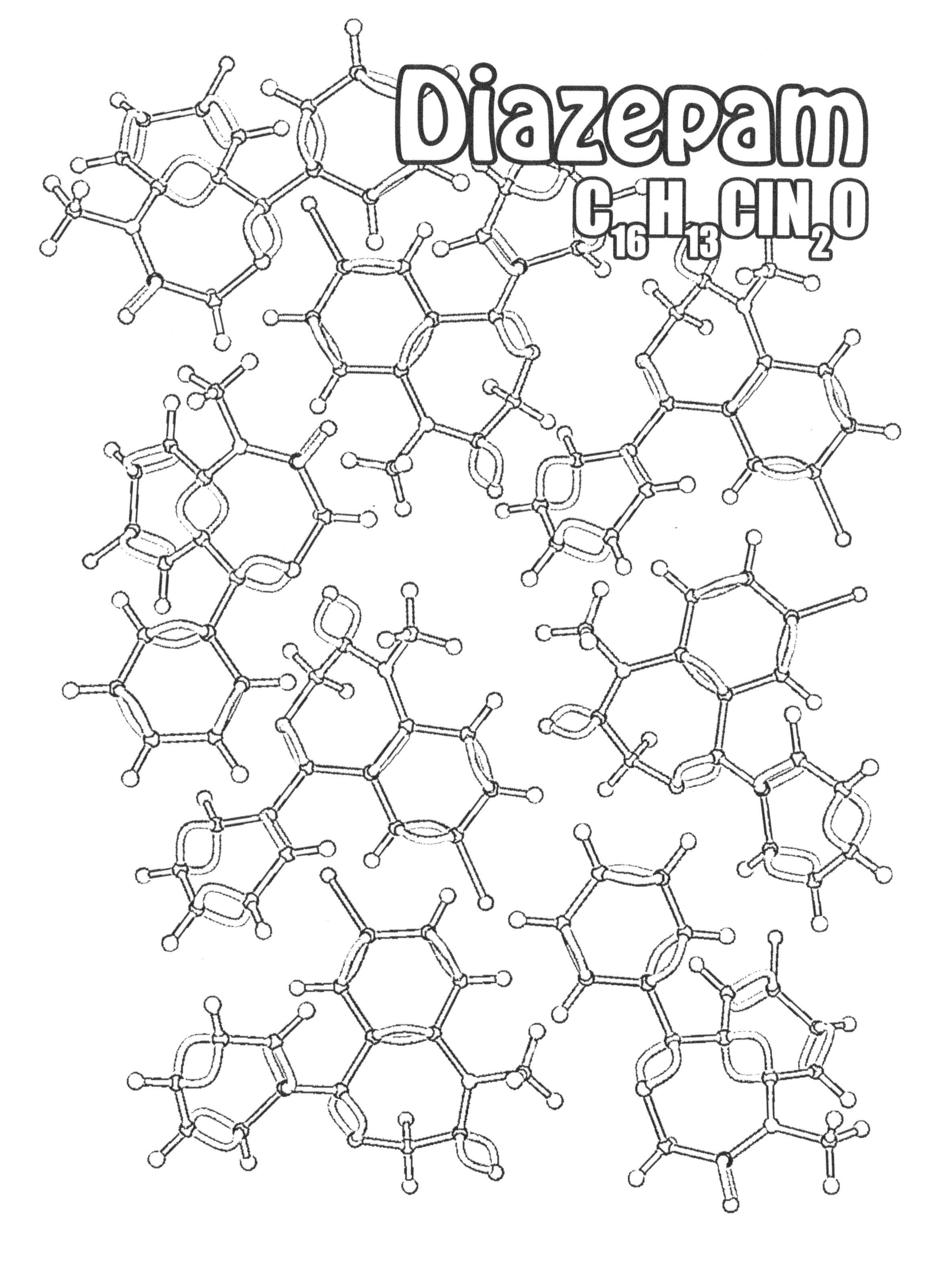
Diazepam
$C_{16}H_{13}ClN_2O$

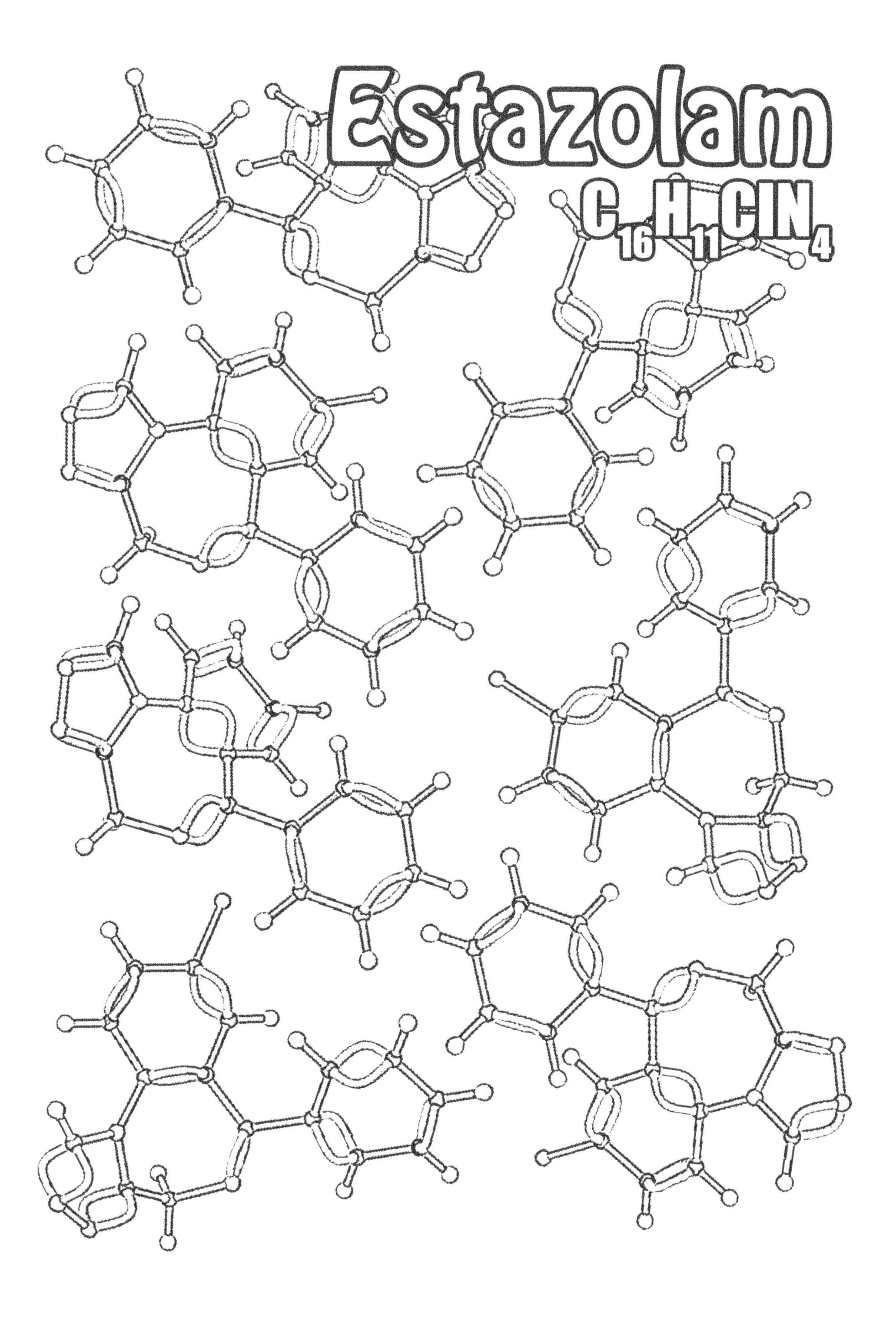

Estazolam
$C_{16}H_{11}ClN_4$

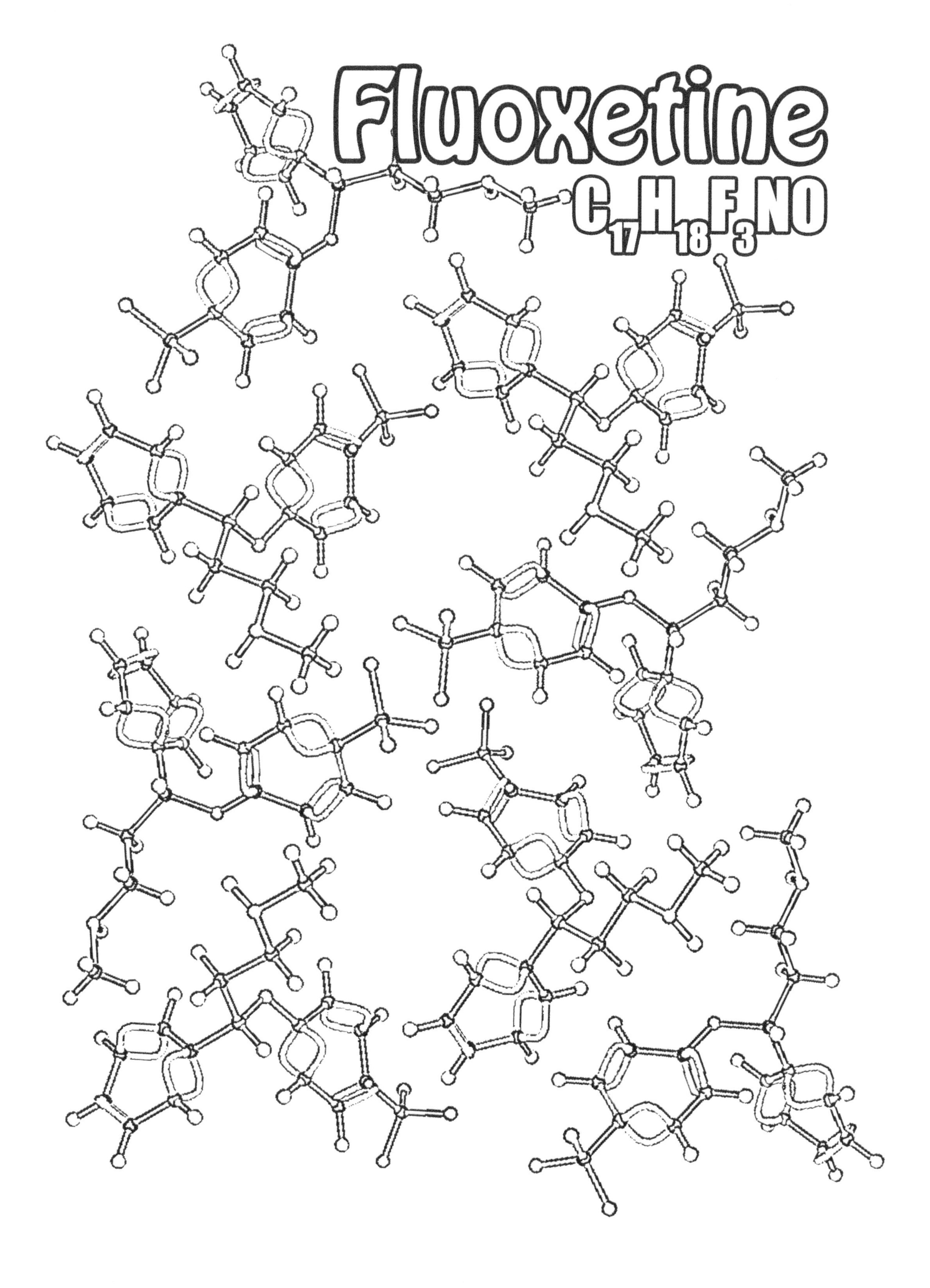
Fluoxetine
$C_{17}H_{18}F_3NO$

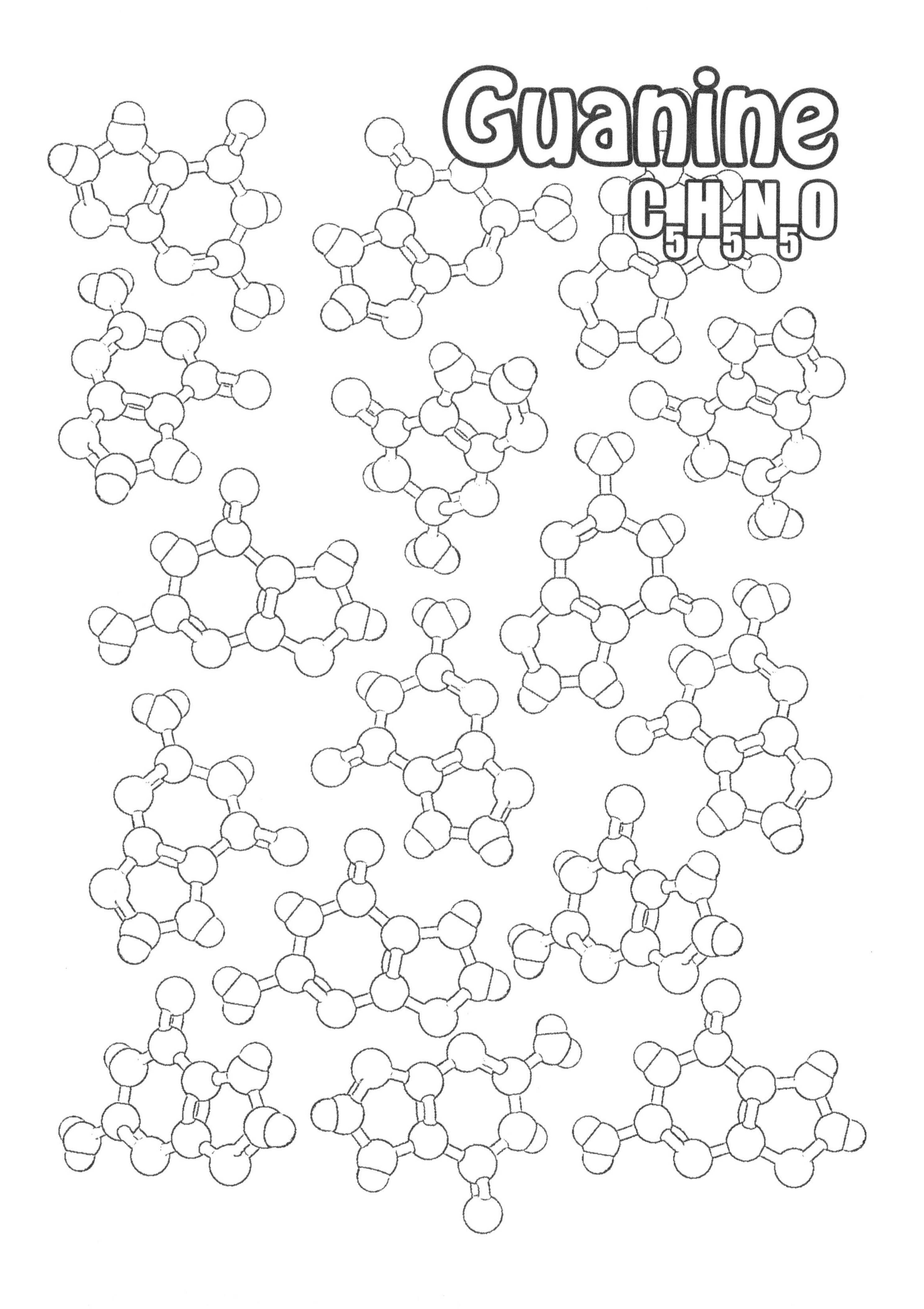
Guanine
C5H5N5O

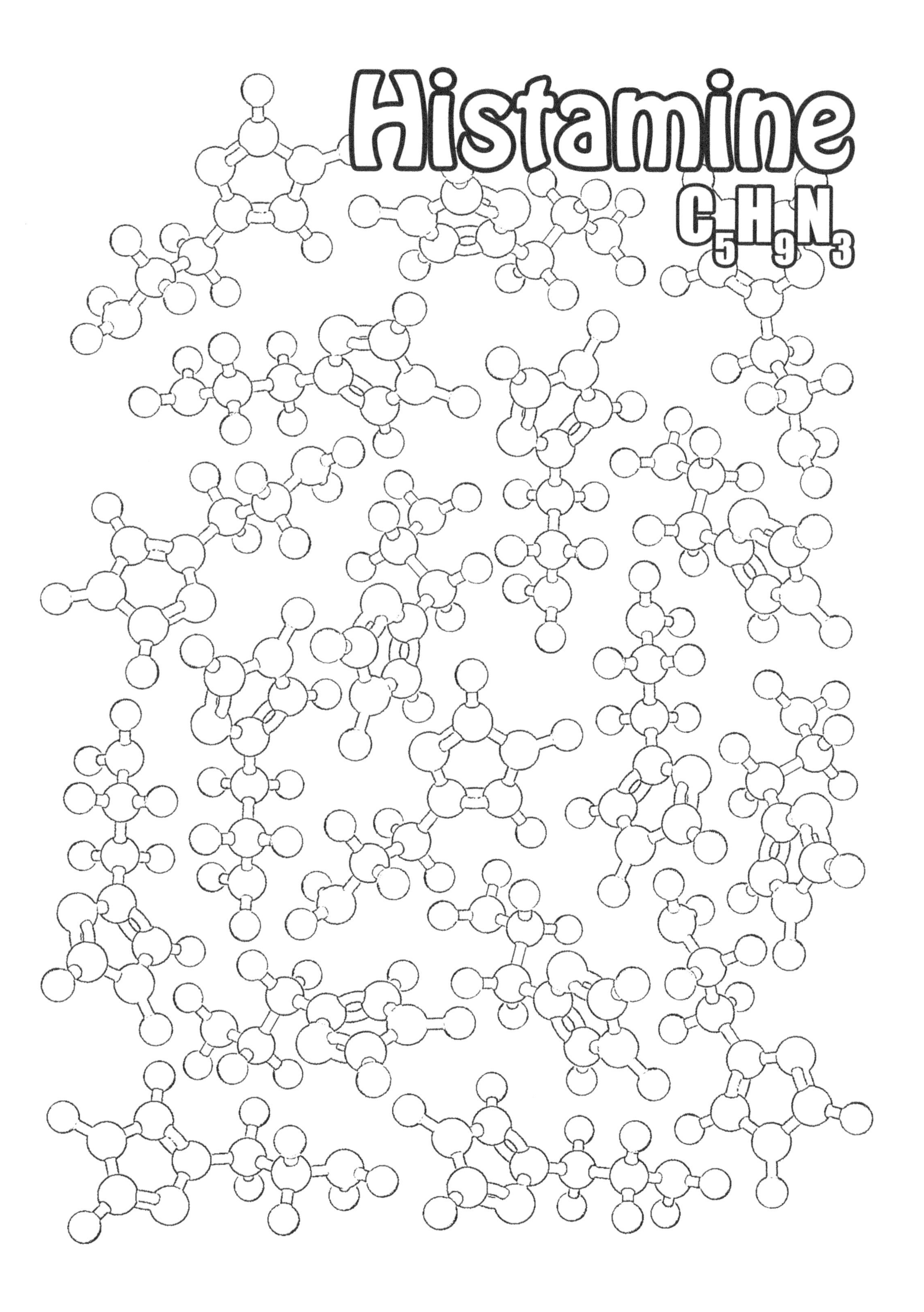
Histamine
$C_5H_9N_3$

Isoprene
C_5H_8

Josamycin
$C_{42}H_{69}NO_{15}$

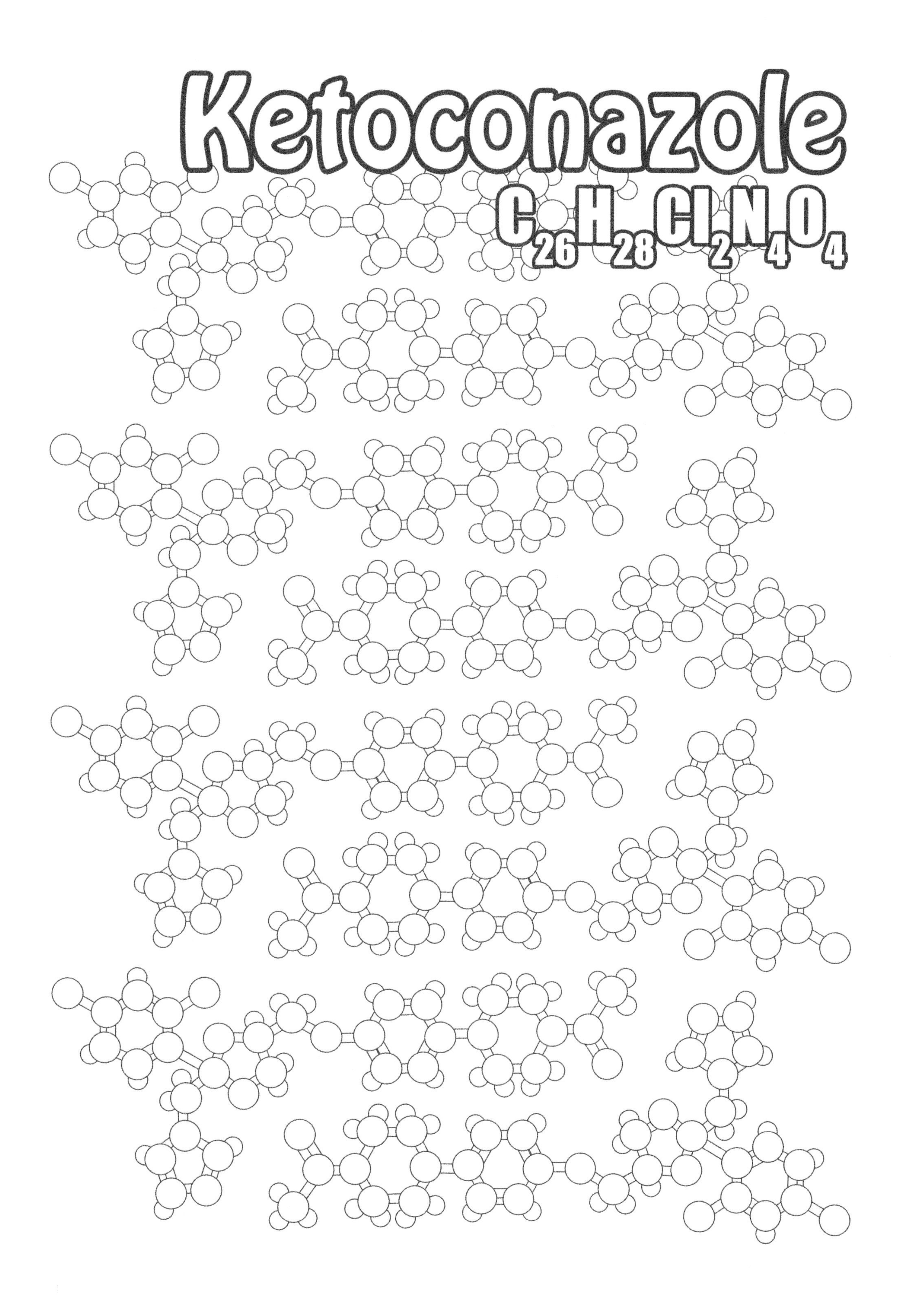
Ketoconazole
$C_{26}H_{28}Cl_{2}N_{4}O_{4}$

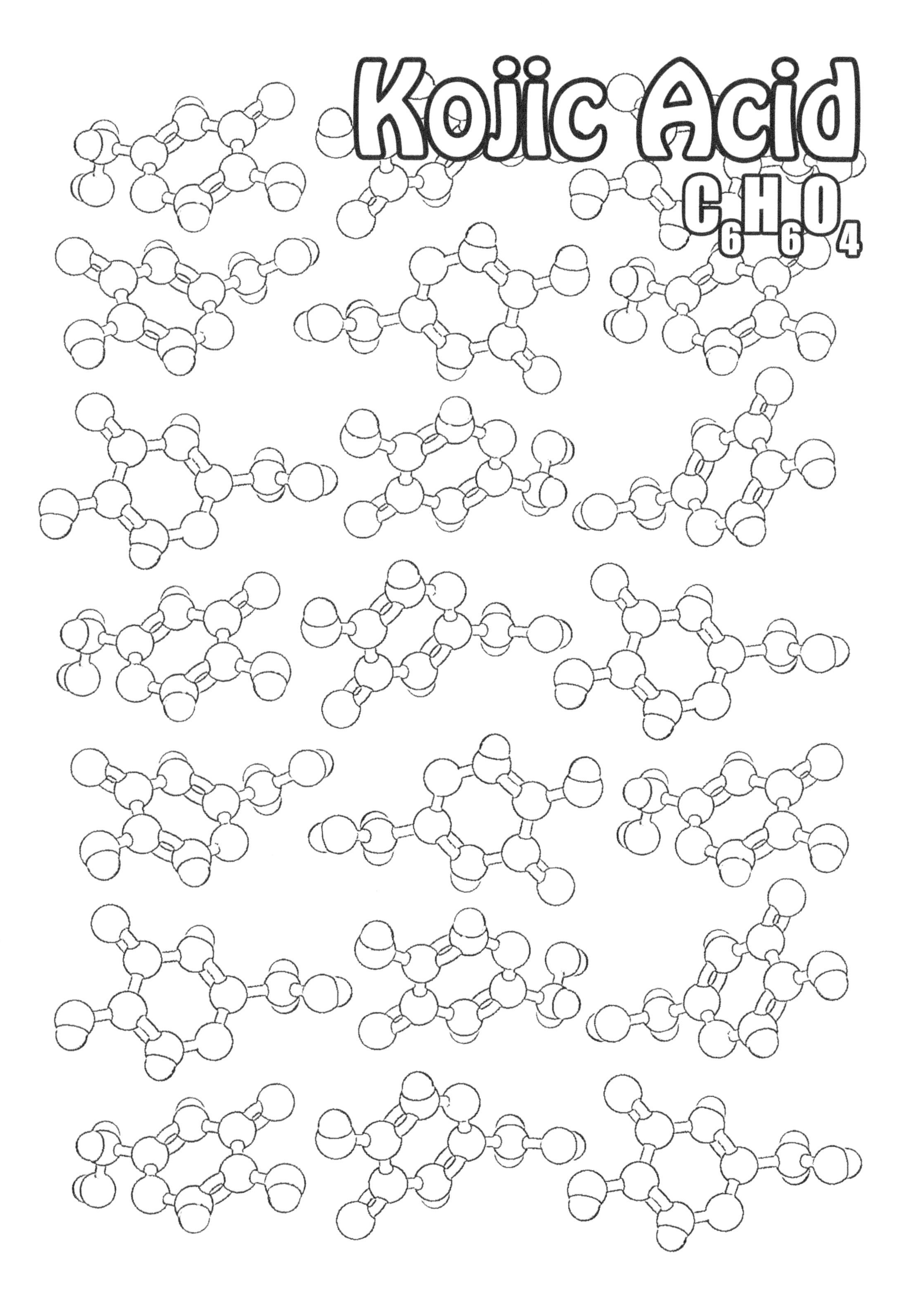
Kojic Acid
$C_6H_6O_4$

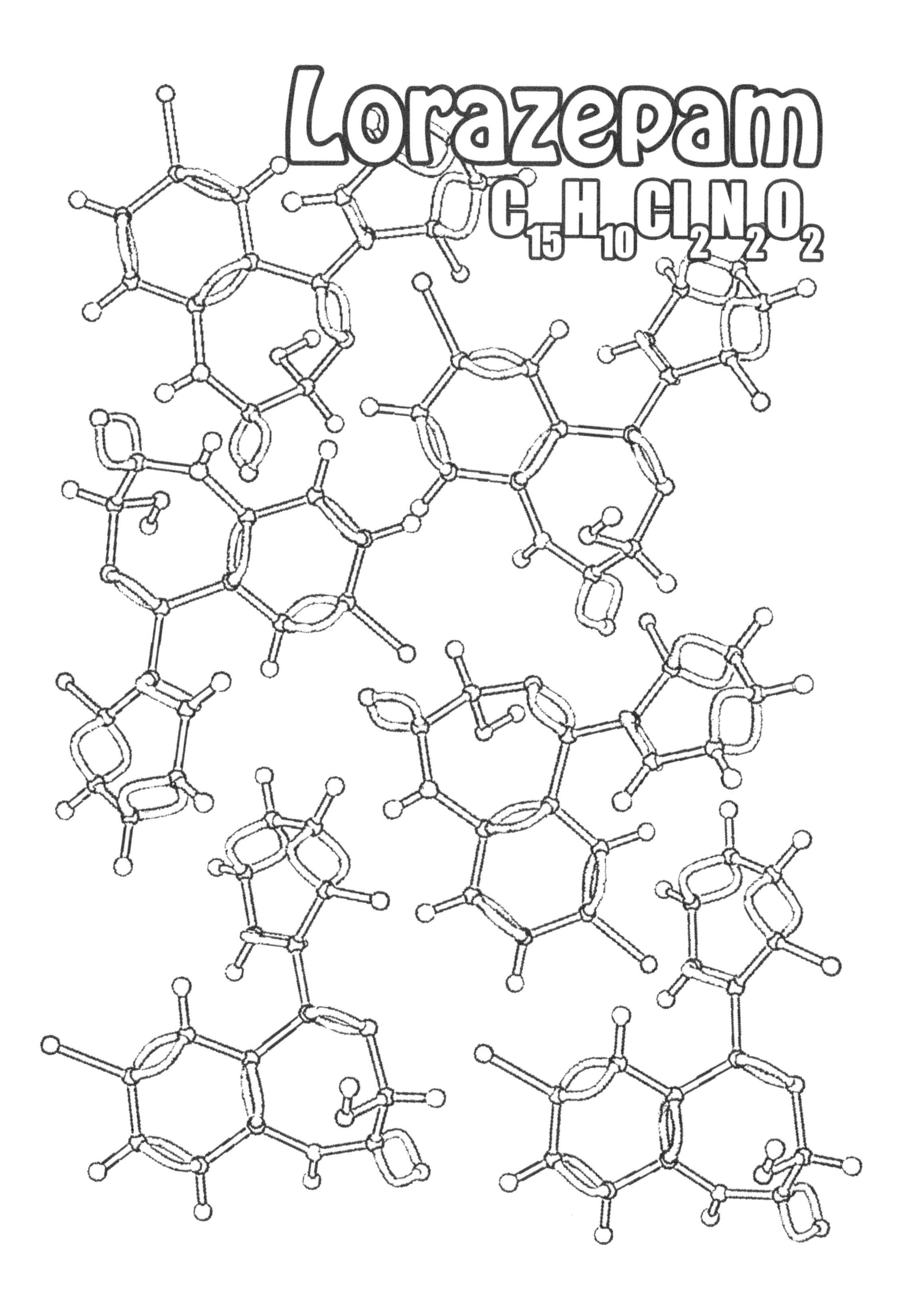
Lorazepam
$C_{15}H_{10}Cl_2N_2O_2$

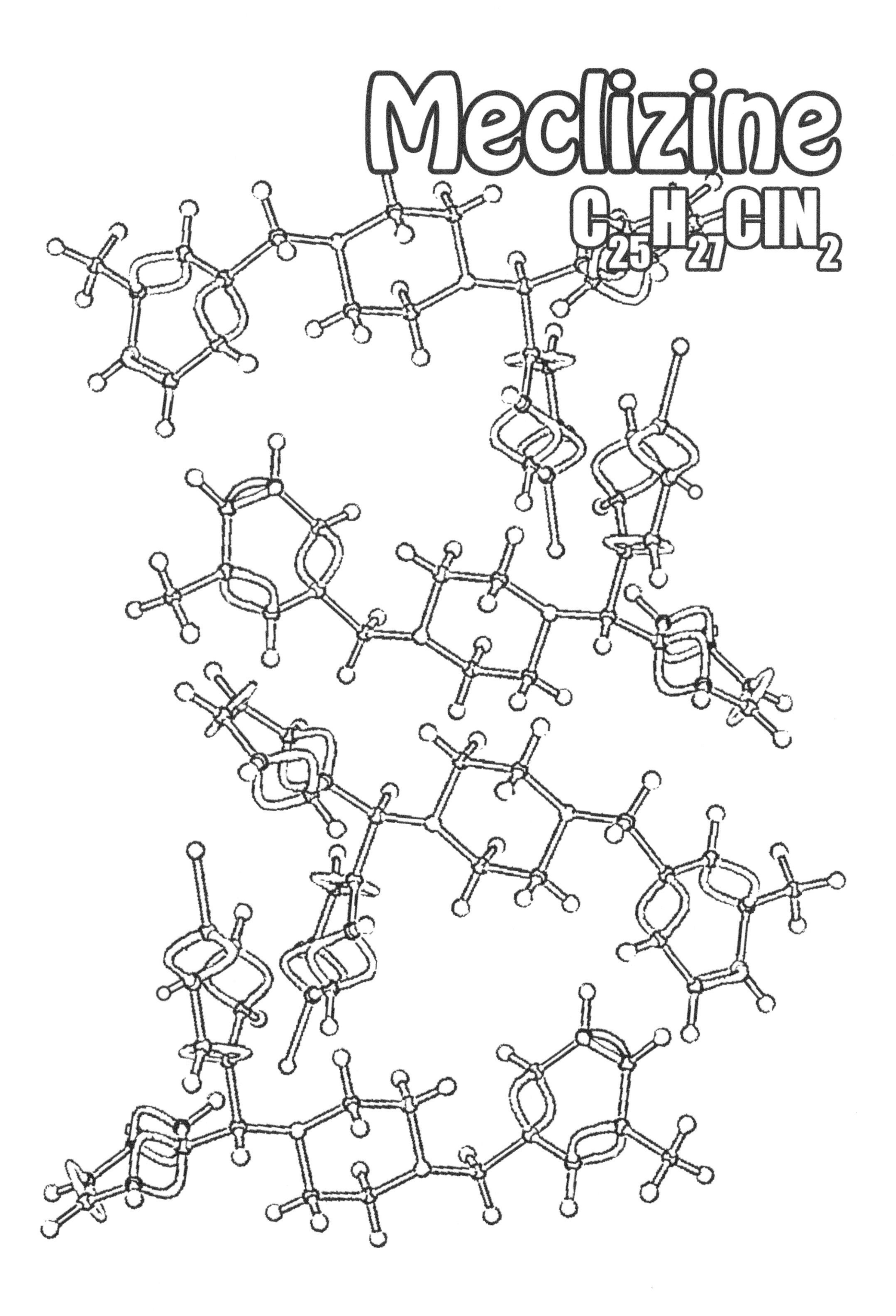
Meclizine
$C_{25}H_{27}ClN_2$

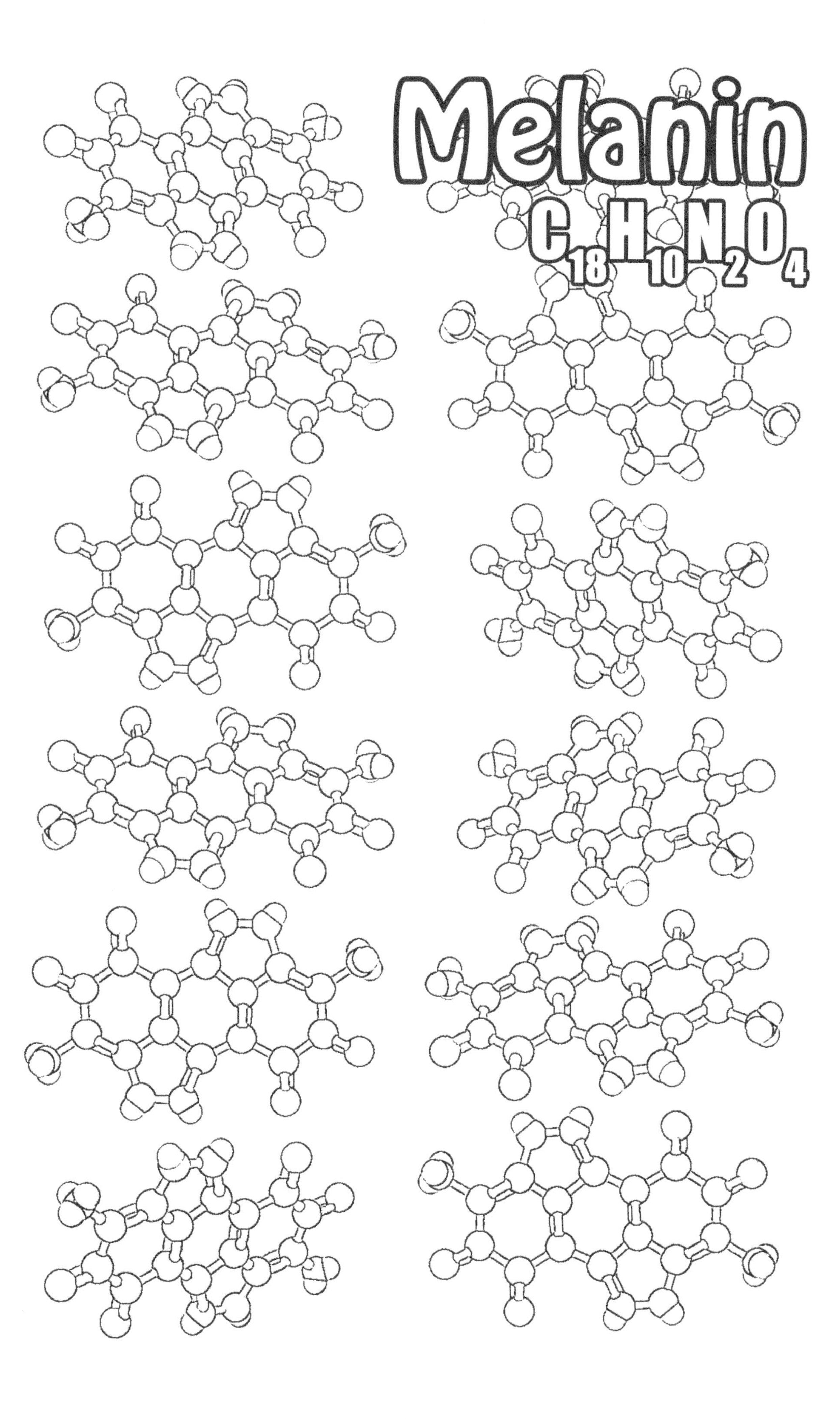
Melanin
$C_{18}H_{10}N_2O_4$

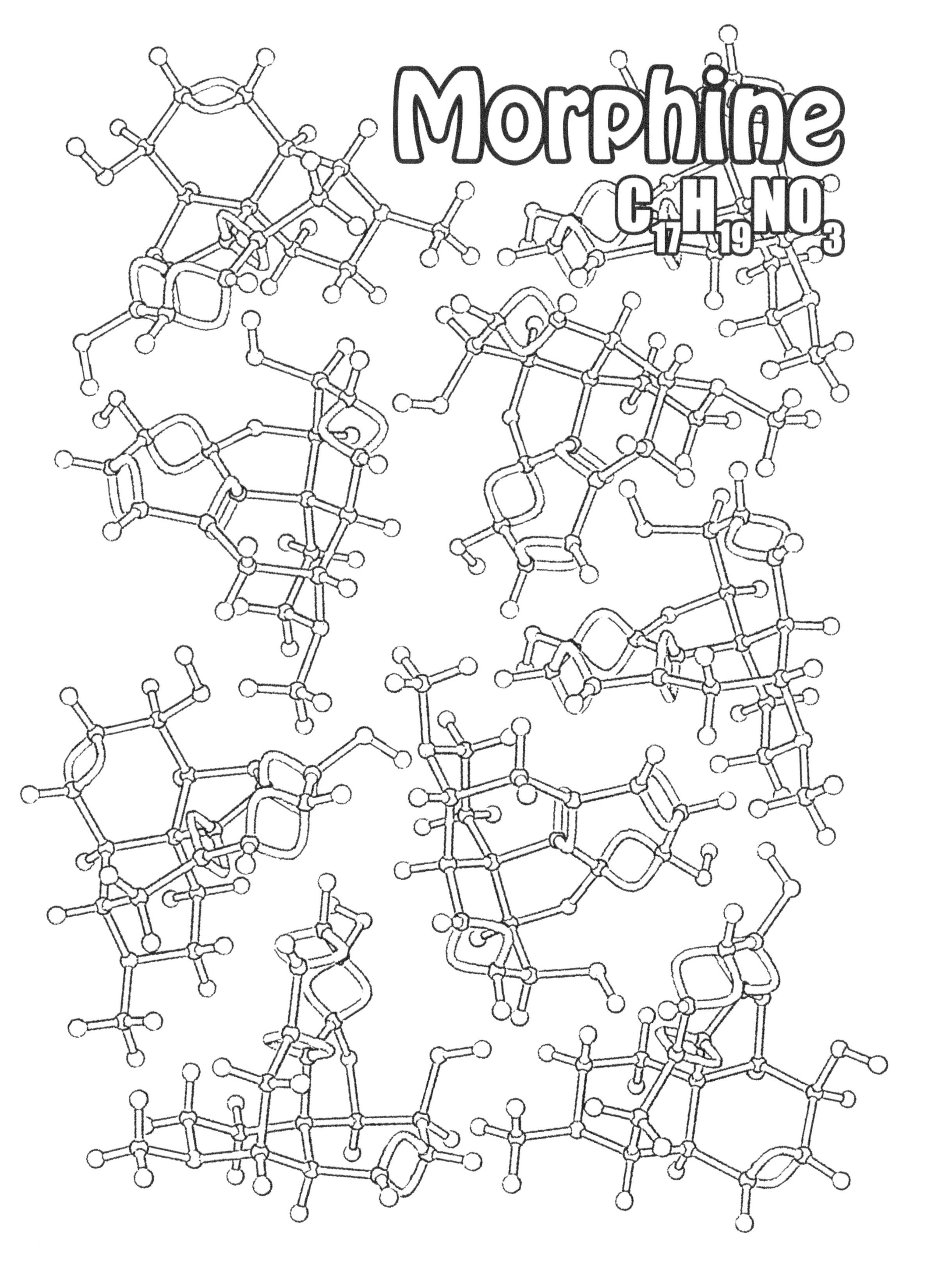
Morphine
$C_{17}H_{19}NO_3$

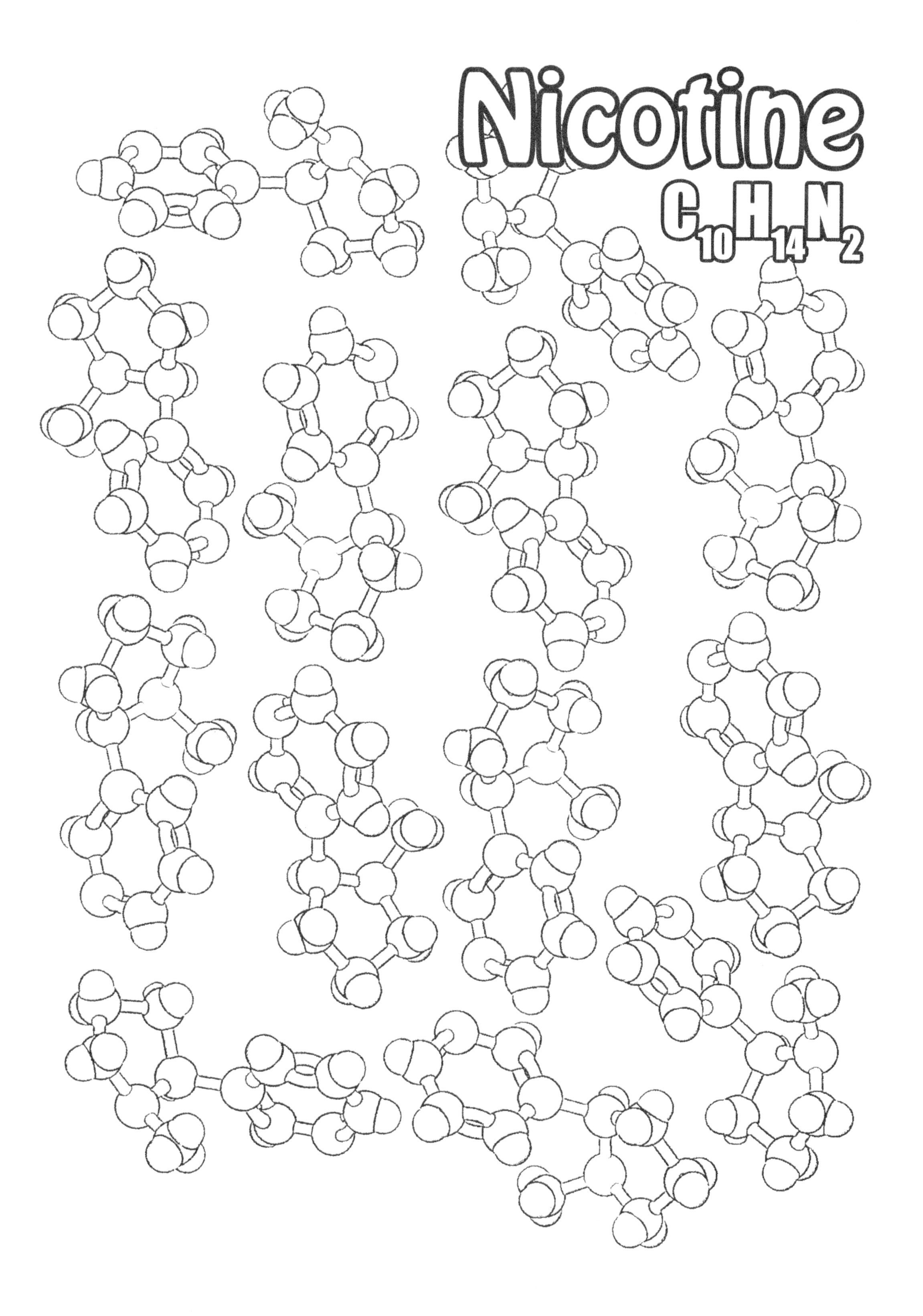
Nicotine
$C_{10}H_{14}N_2$

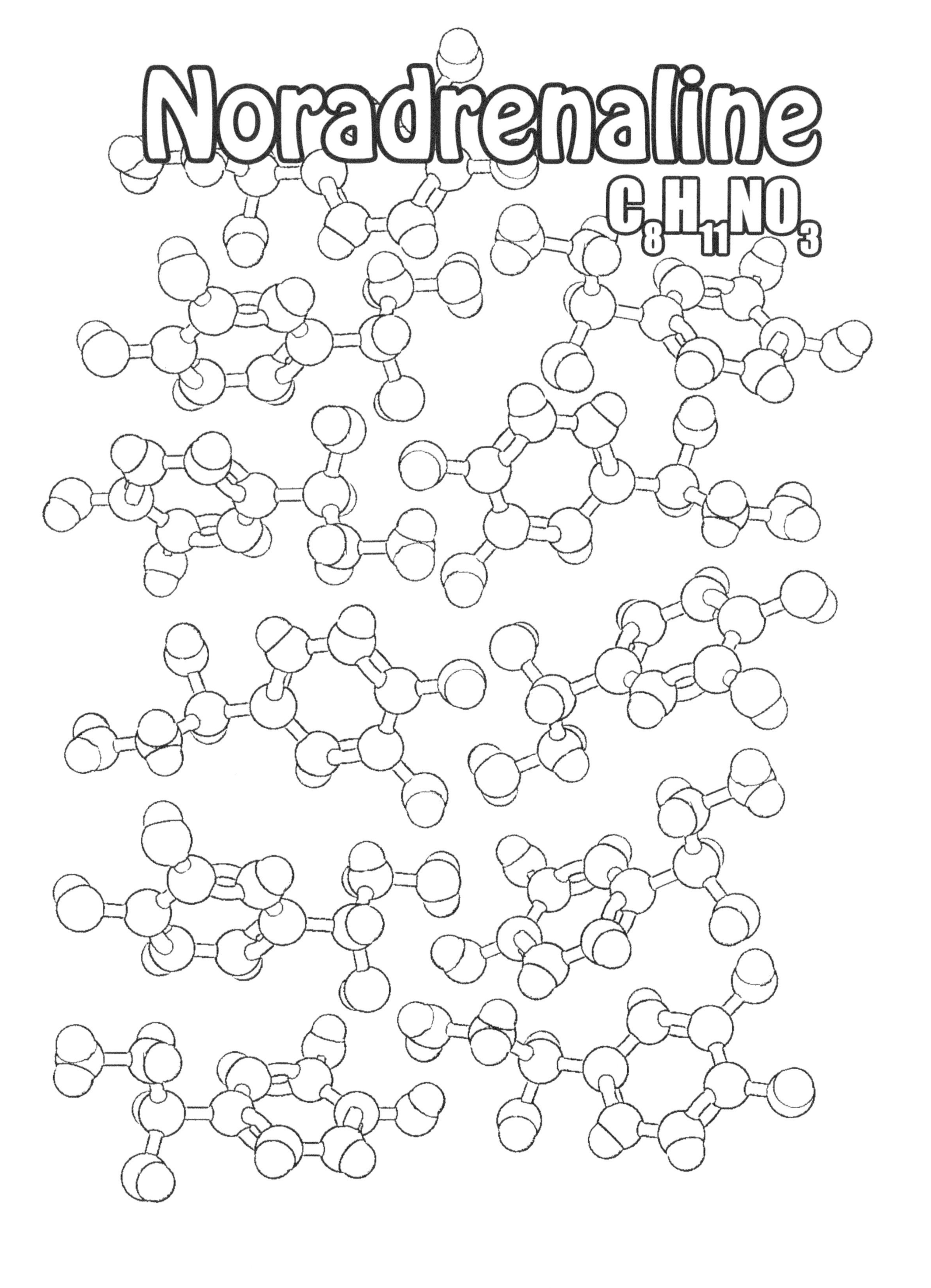
Noradrenaline
C8H11NO3

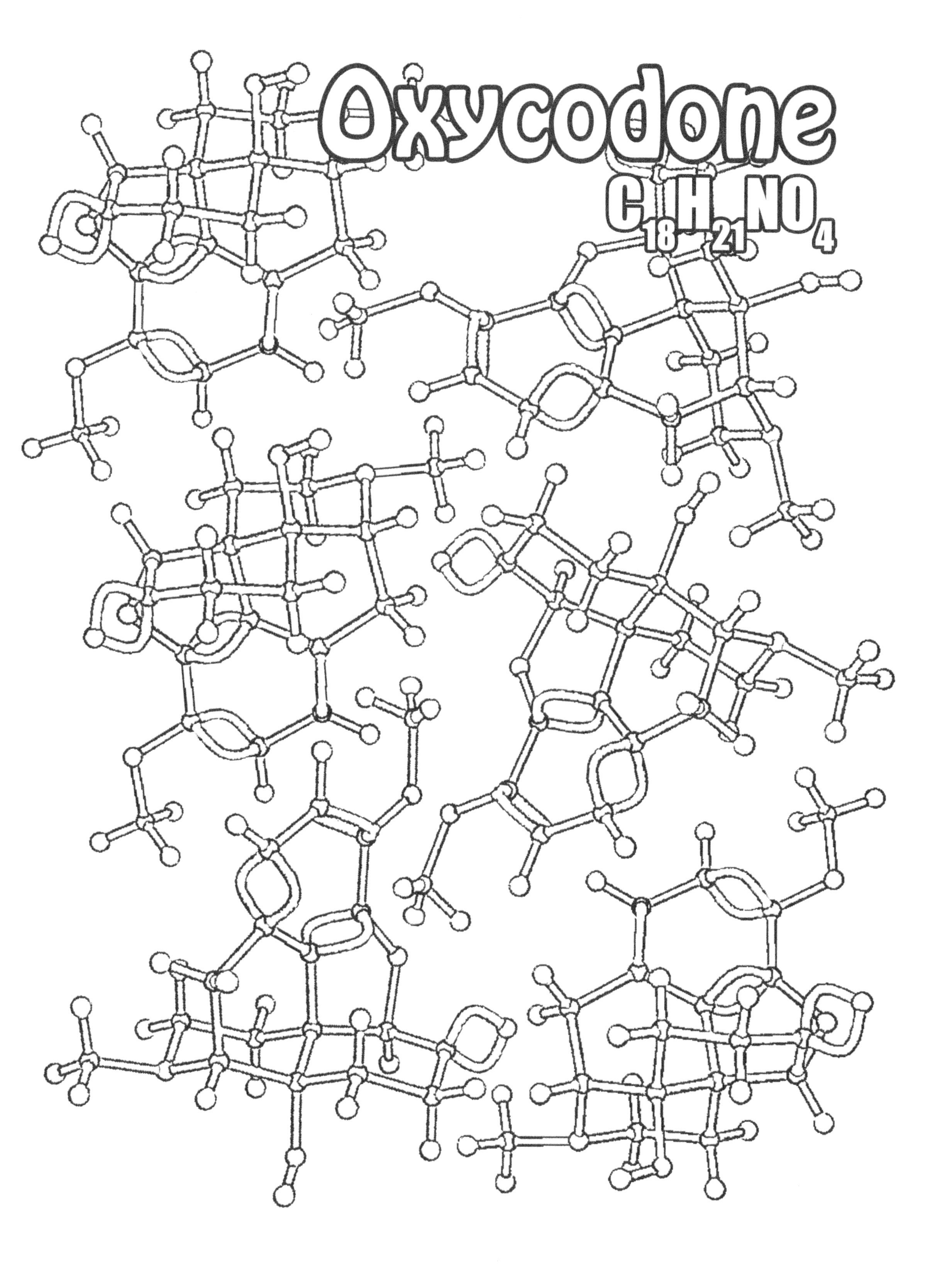
Oxycodone
$C_{18}H_{21}NO_4$

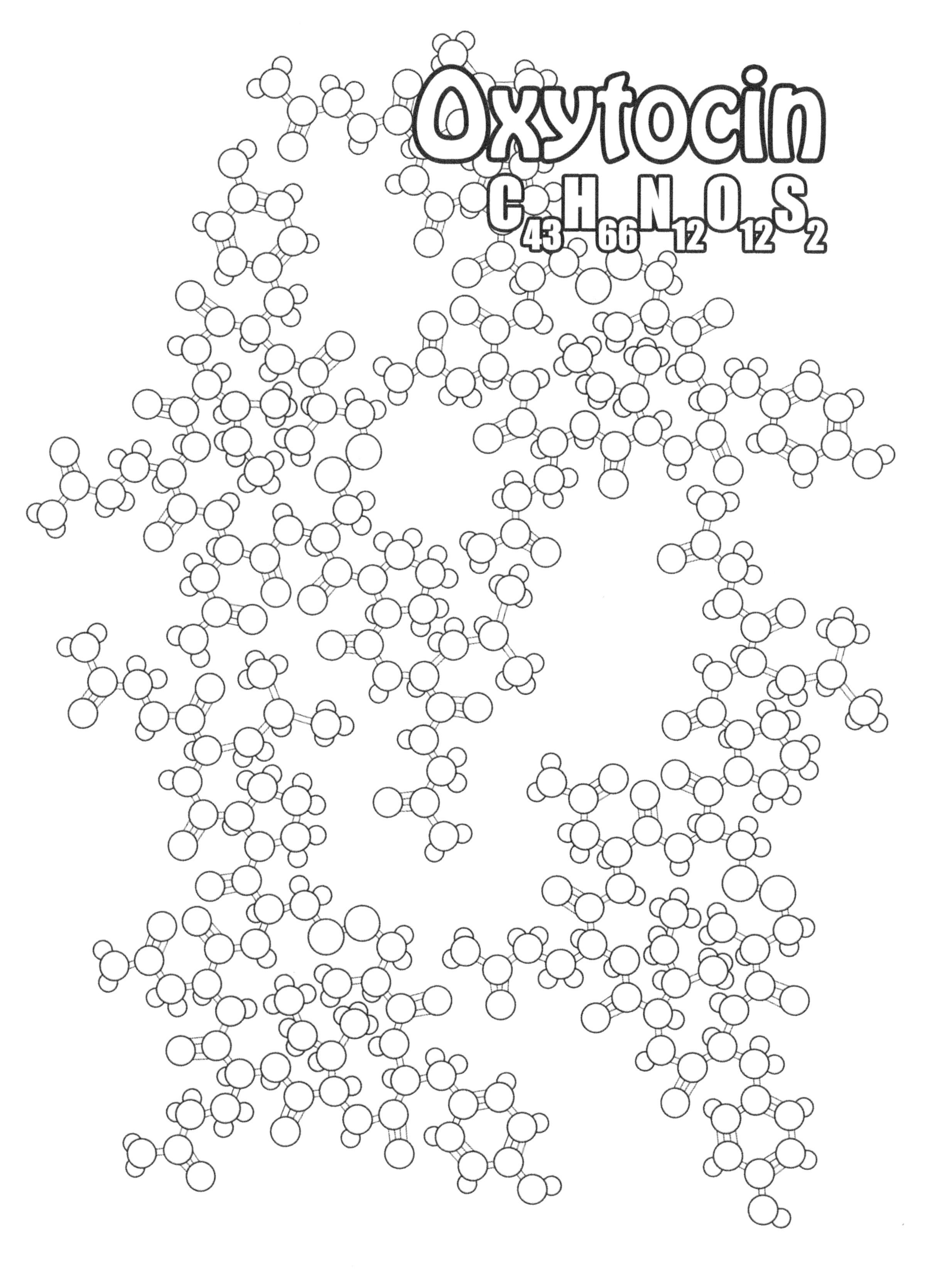
Oxytocin
$C_{43}H_{66}N_{12}O_{12}S_2$

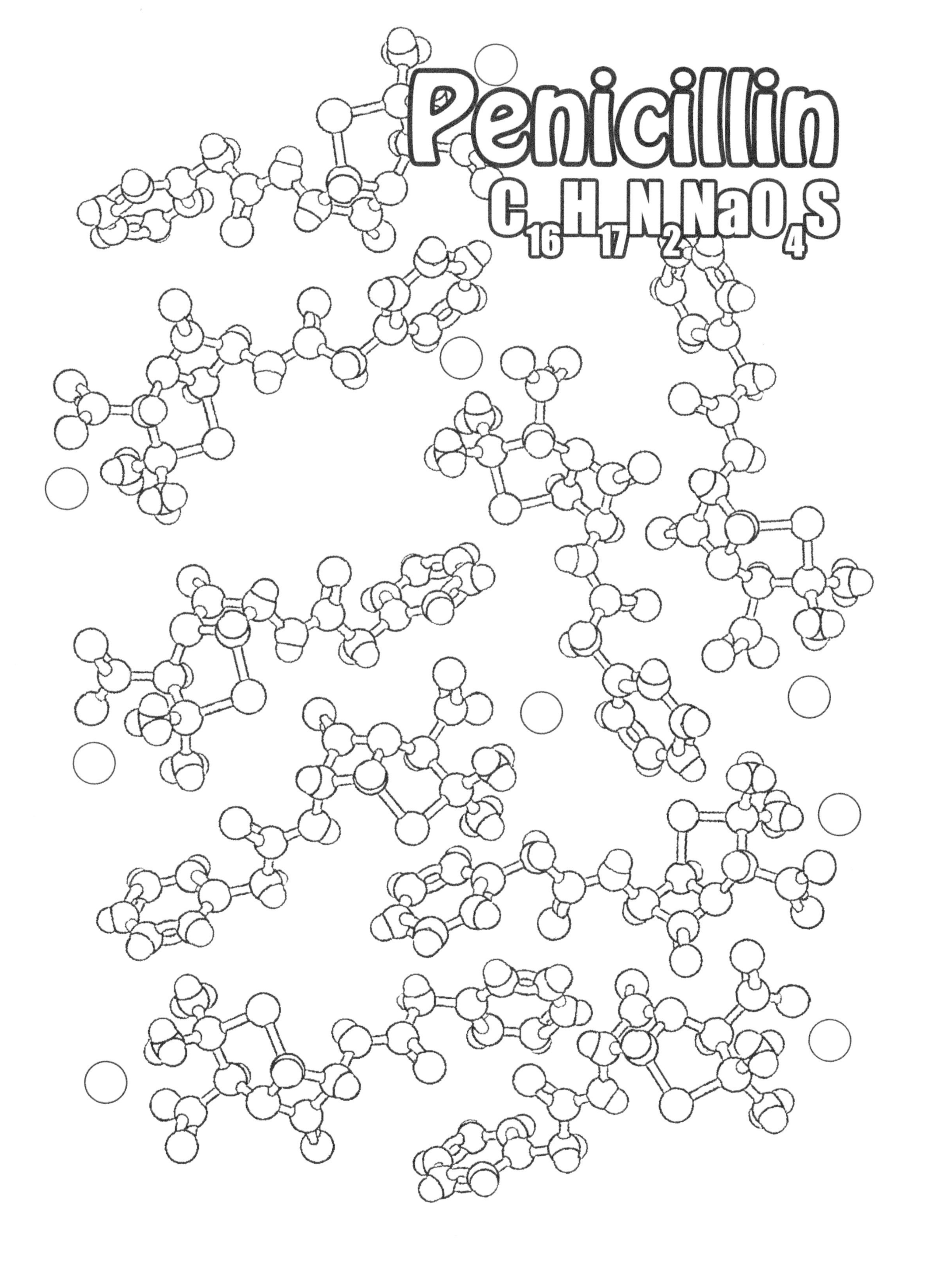
Penicillin
$C_{16}H_{17}N_{2}NaO_{4}S$

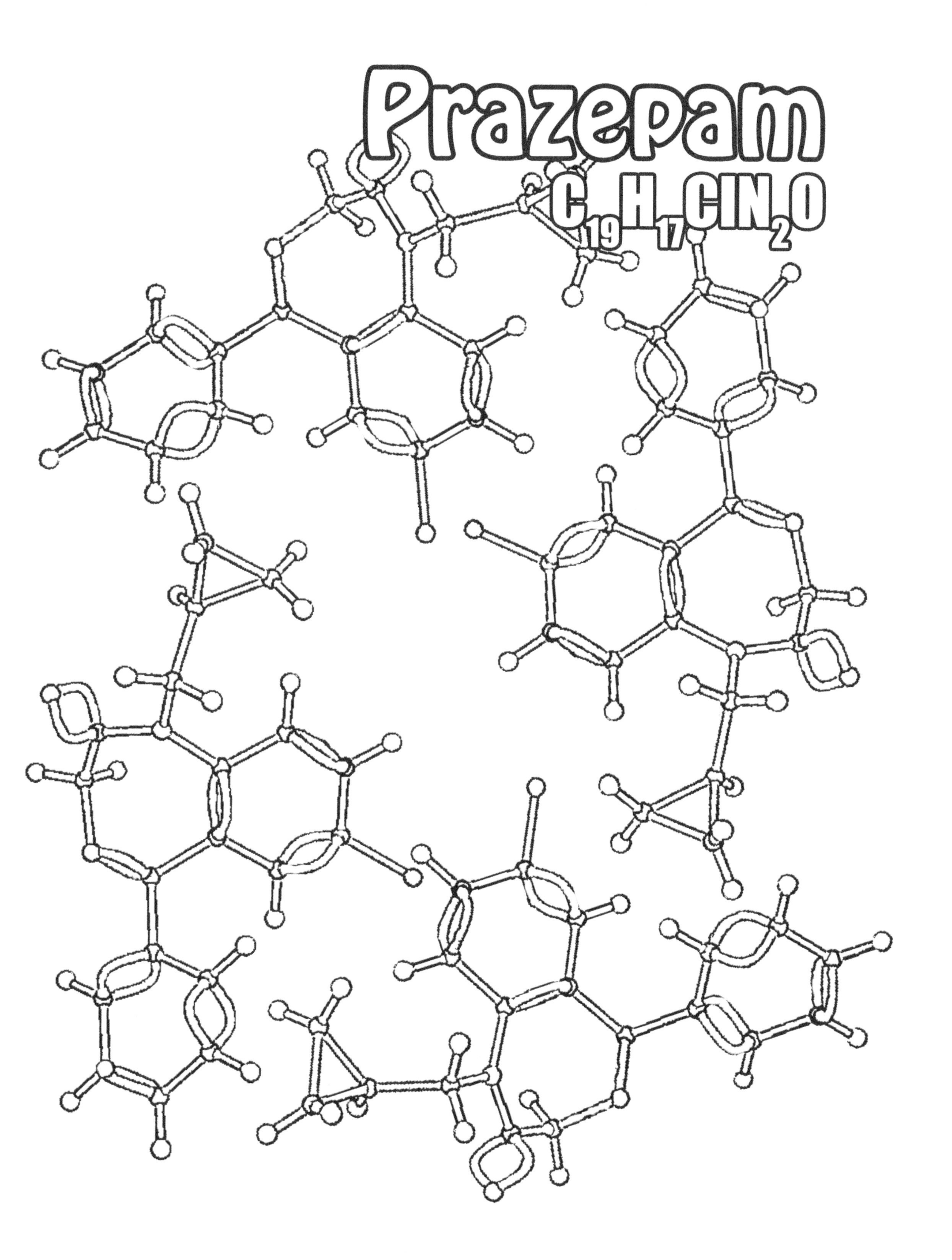
Prazepam
$C_{19}H_{17}ClN_2O$

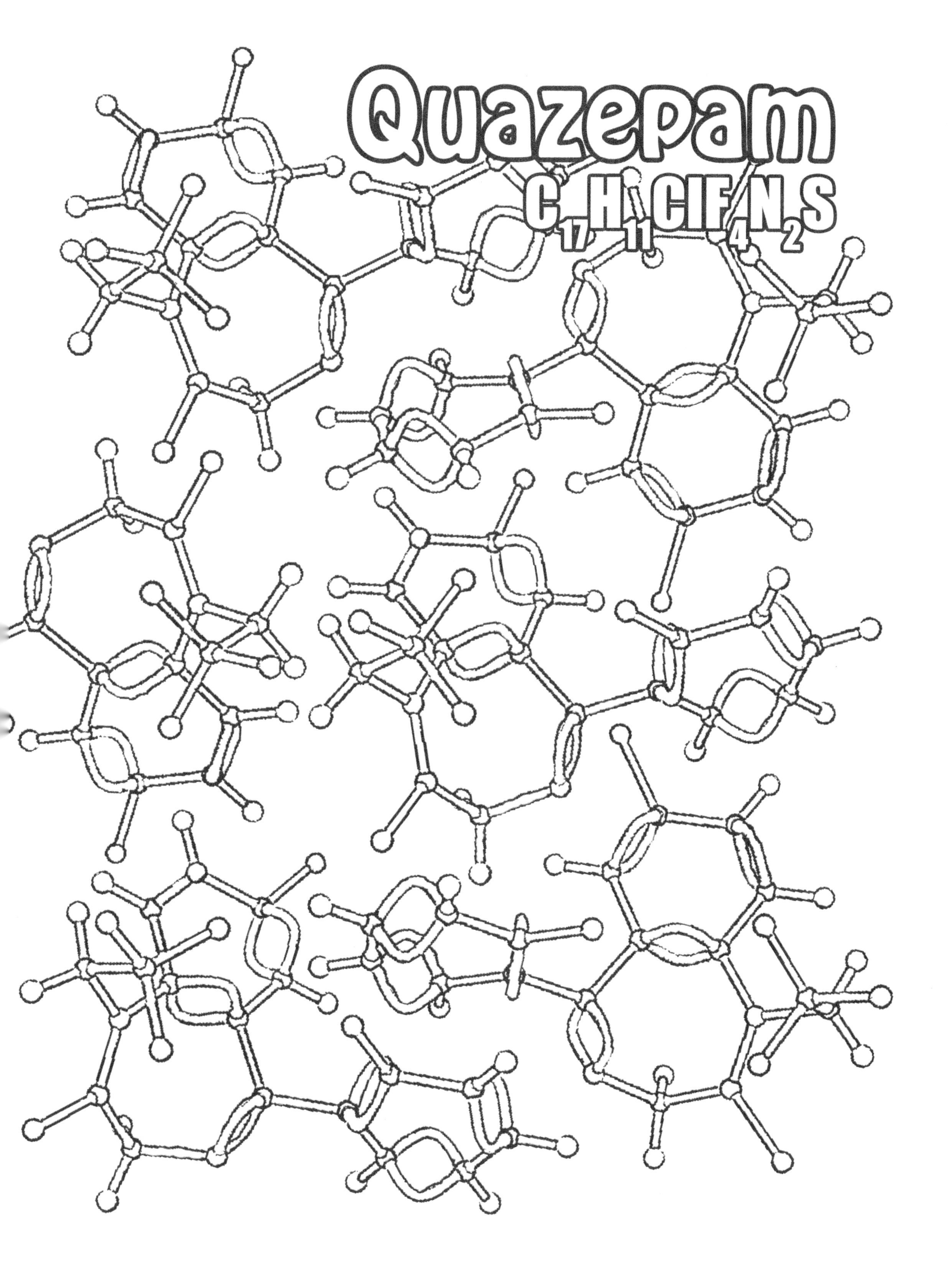
Quazepam
C17H11ClF4N2S

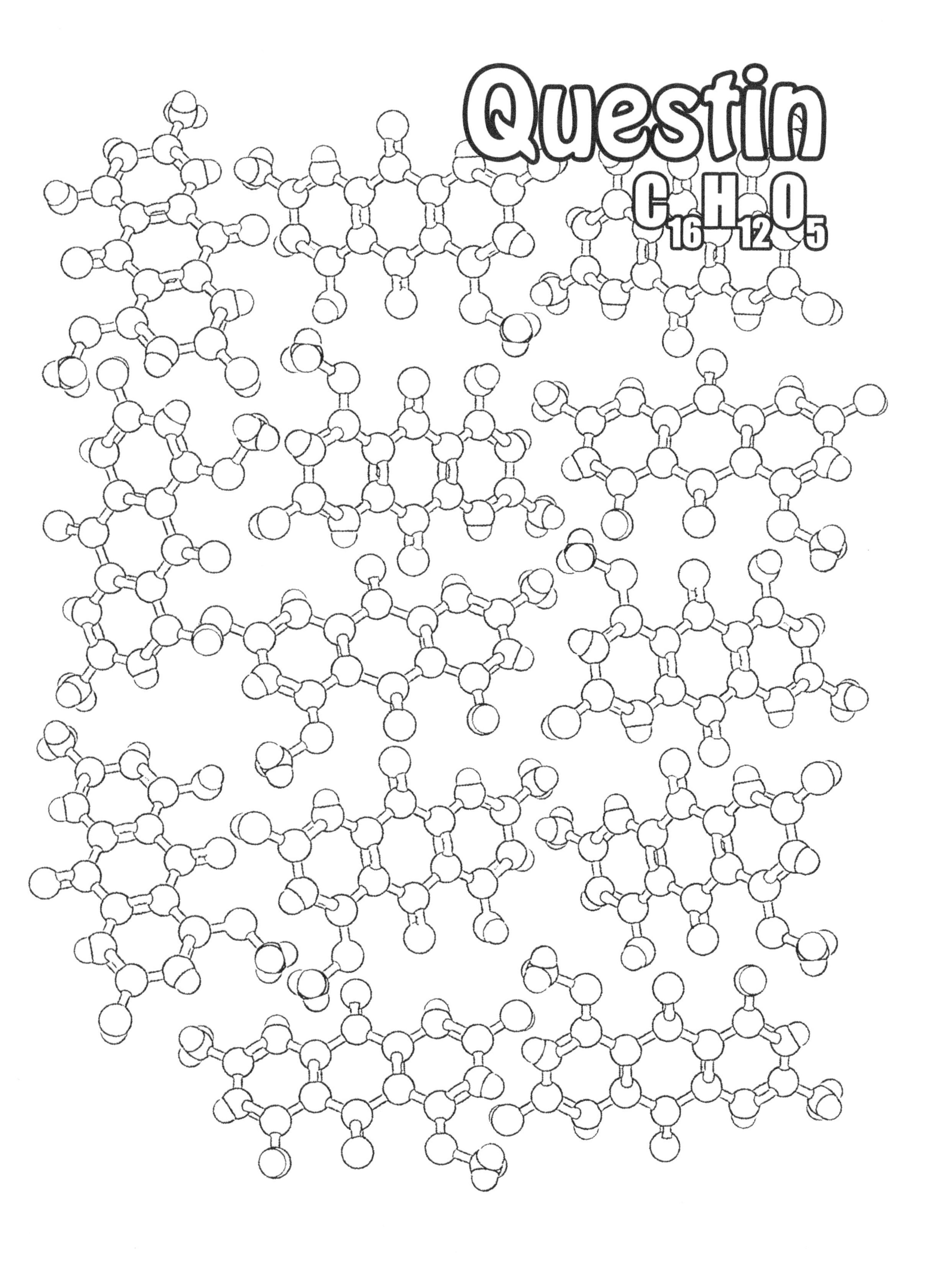
Questin
$C_{16}H_{12}O_5$

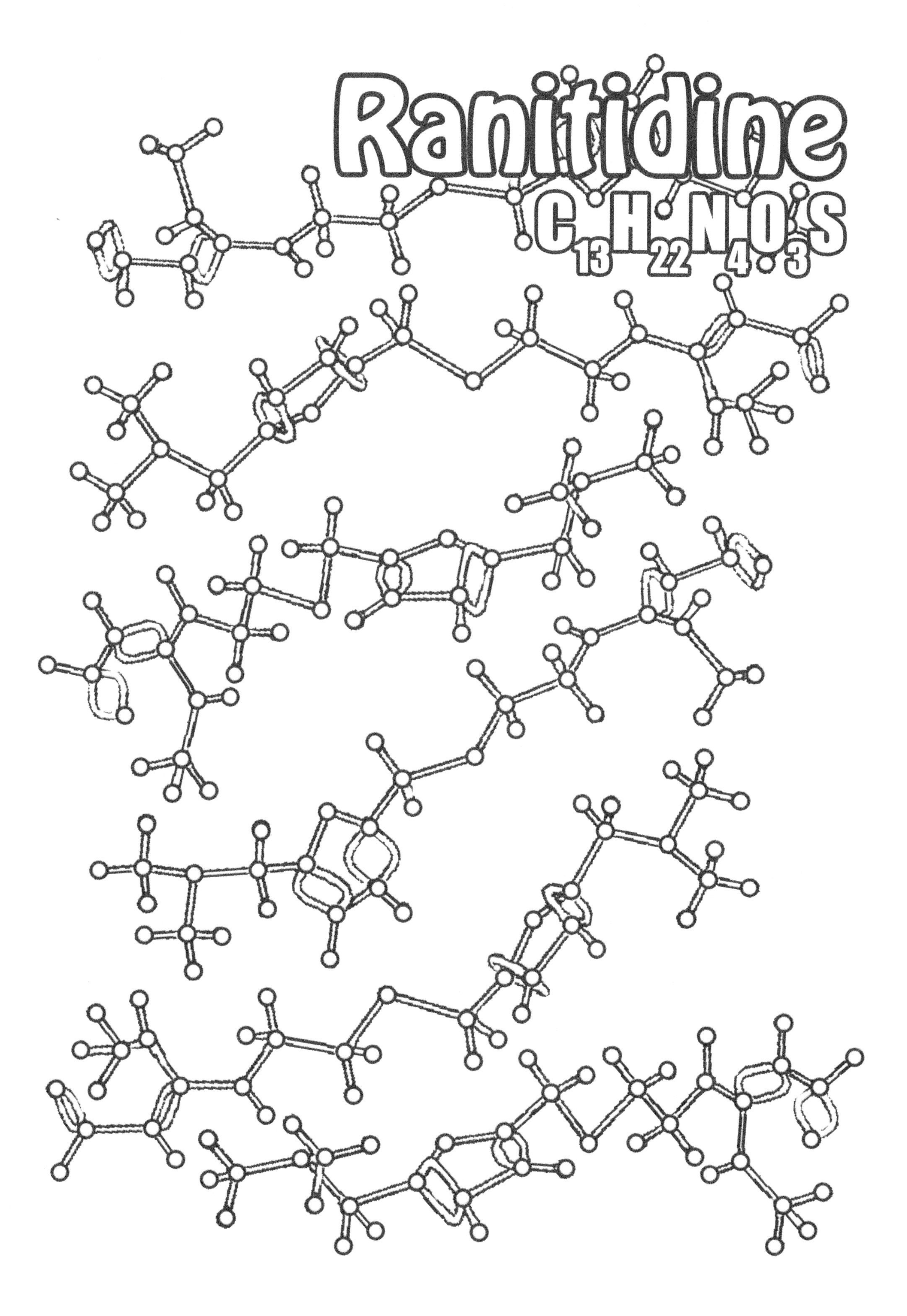
Ranitidine
C13H22N4O3S

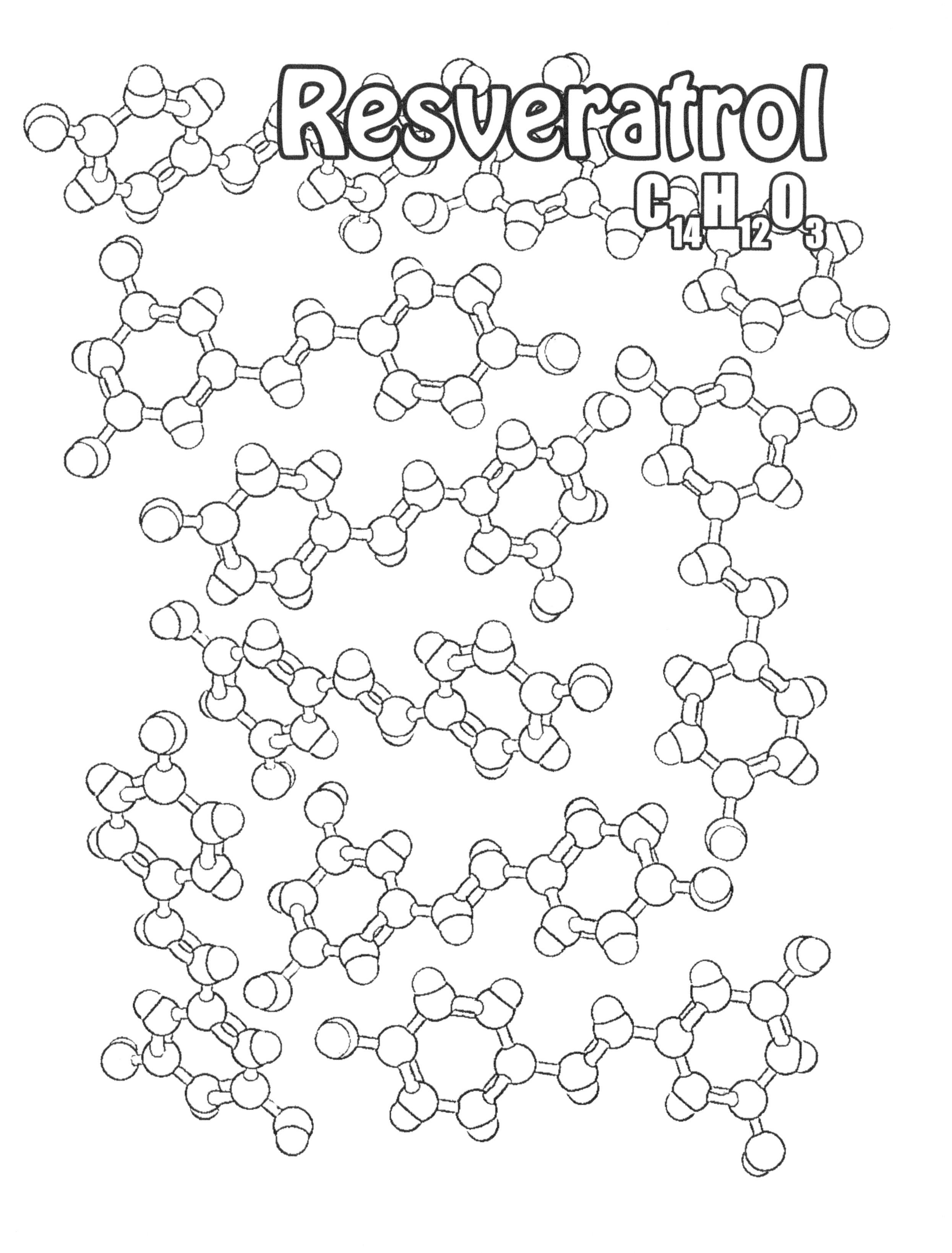

Resveratrol
$C_{14}H_{12}O_3$

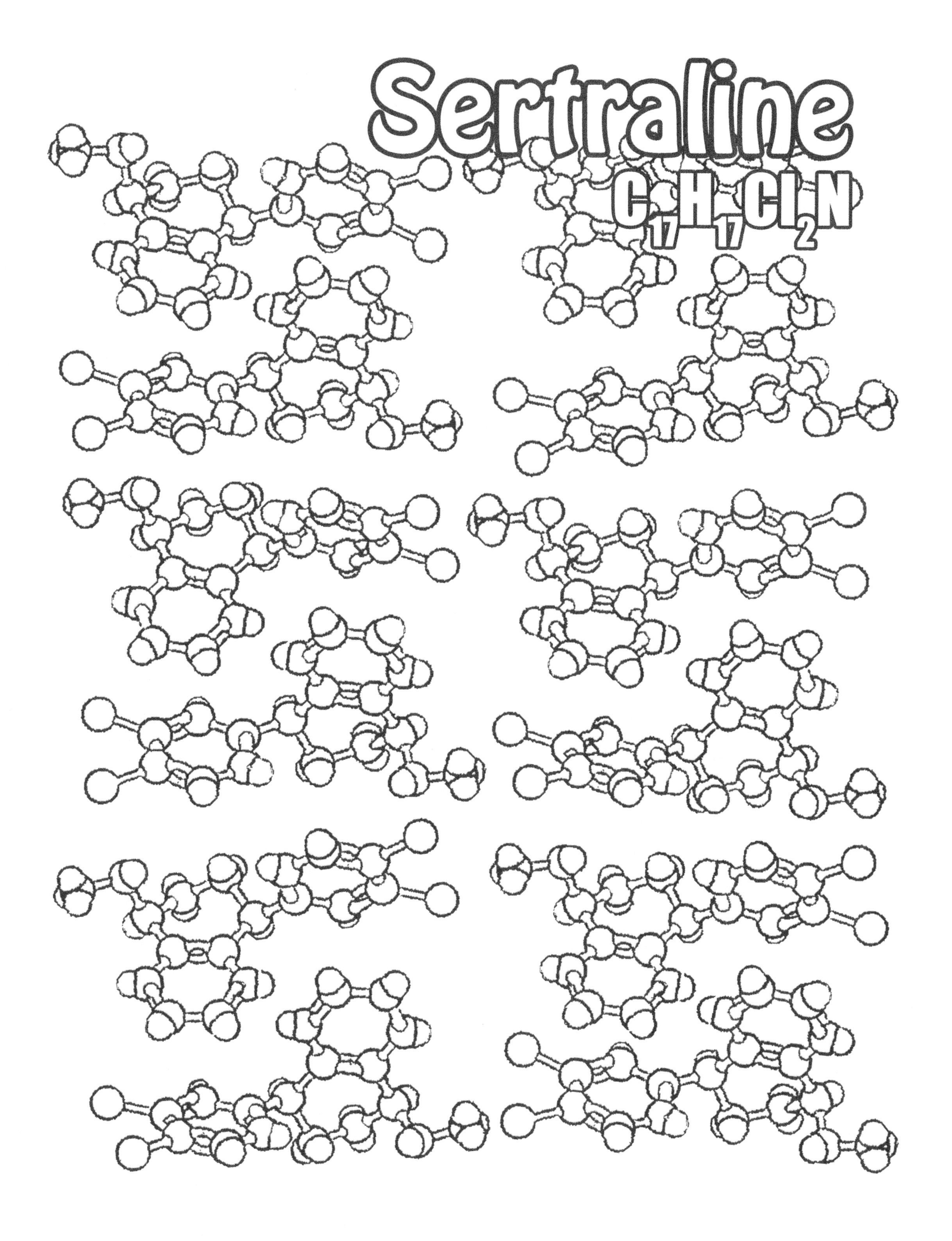
Sertraline
$C_{17}H_{17}Cl_2N$

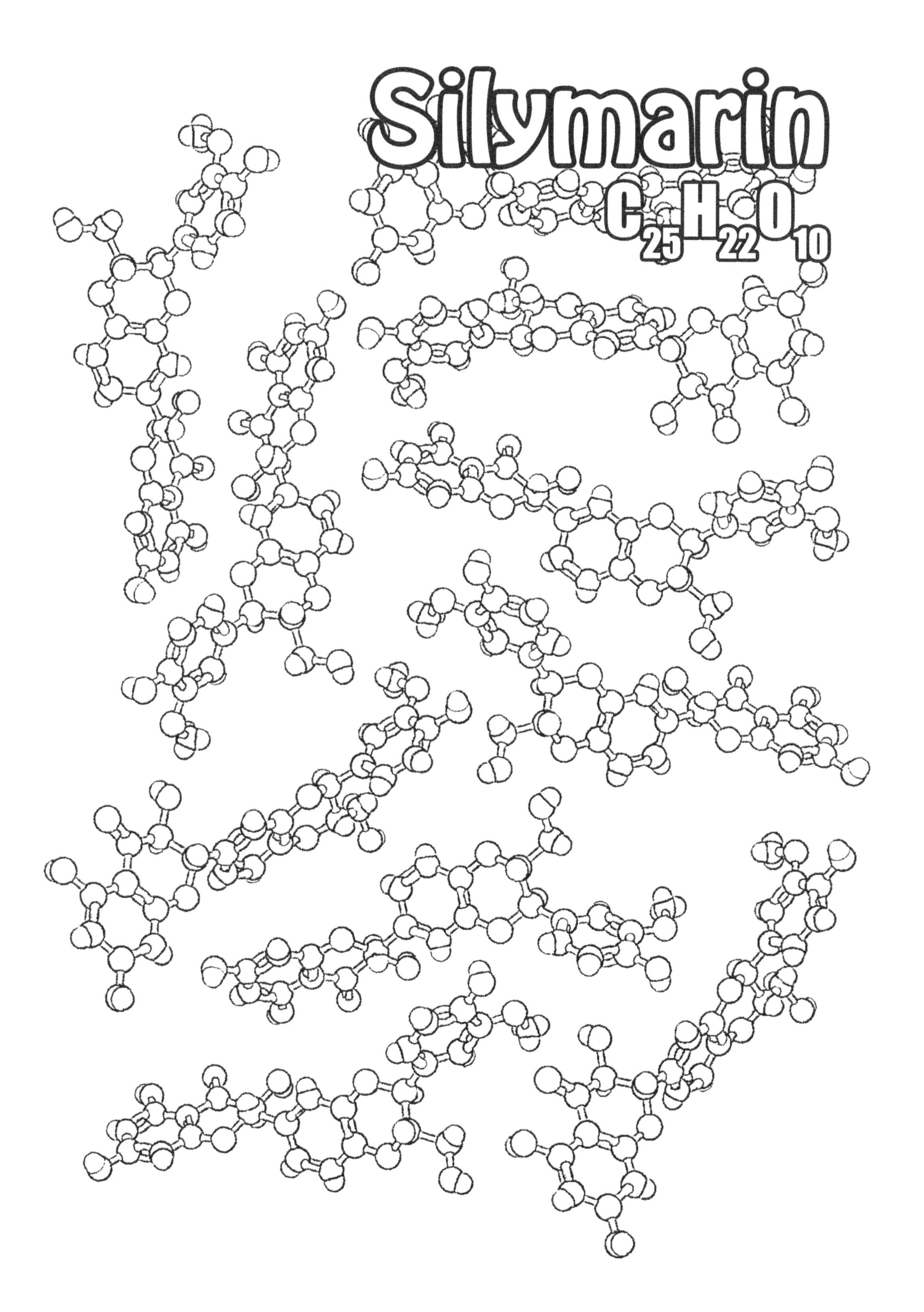
Silymarin
$C_{25}H_{22}O_{10}$

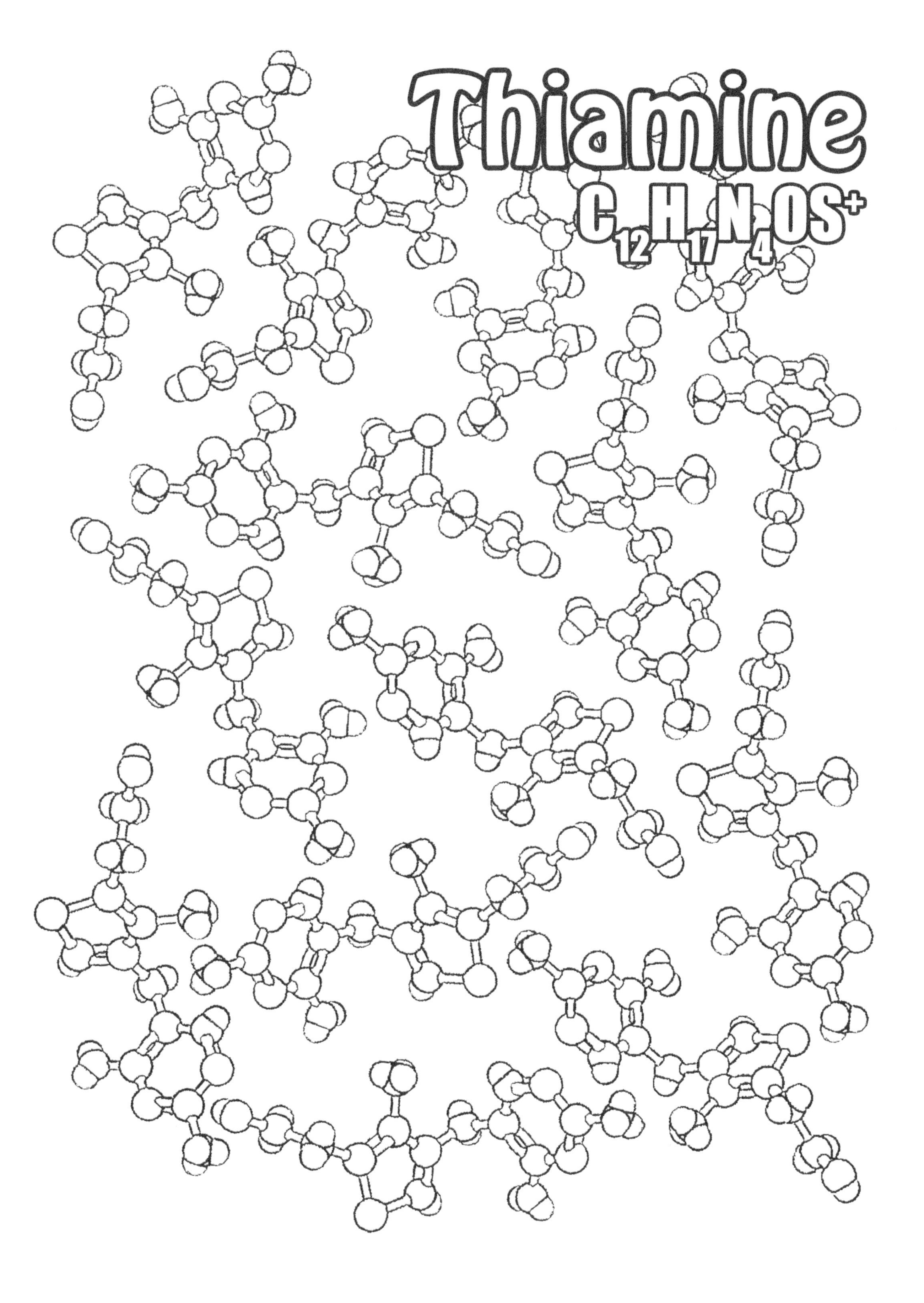
Thiamine
$C_{12}H_{17}N_4OS^+$

Thymine

$C_5H_6N_2O_2$

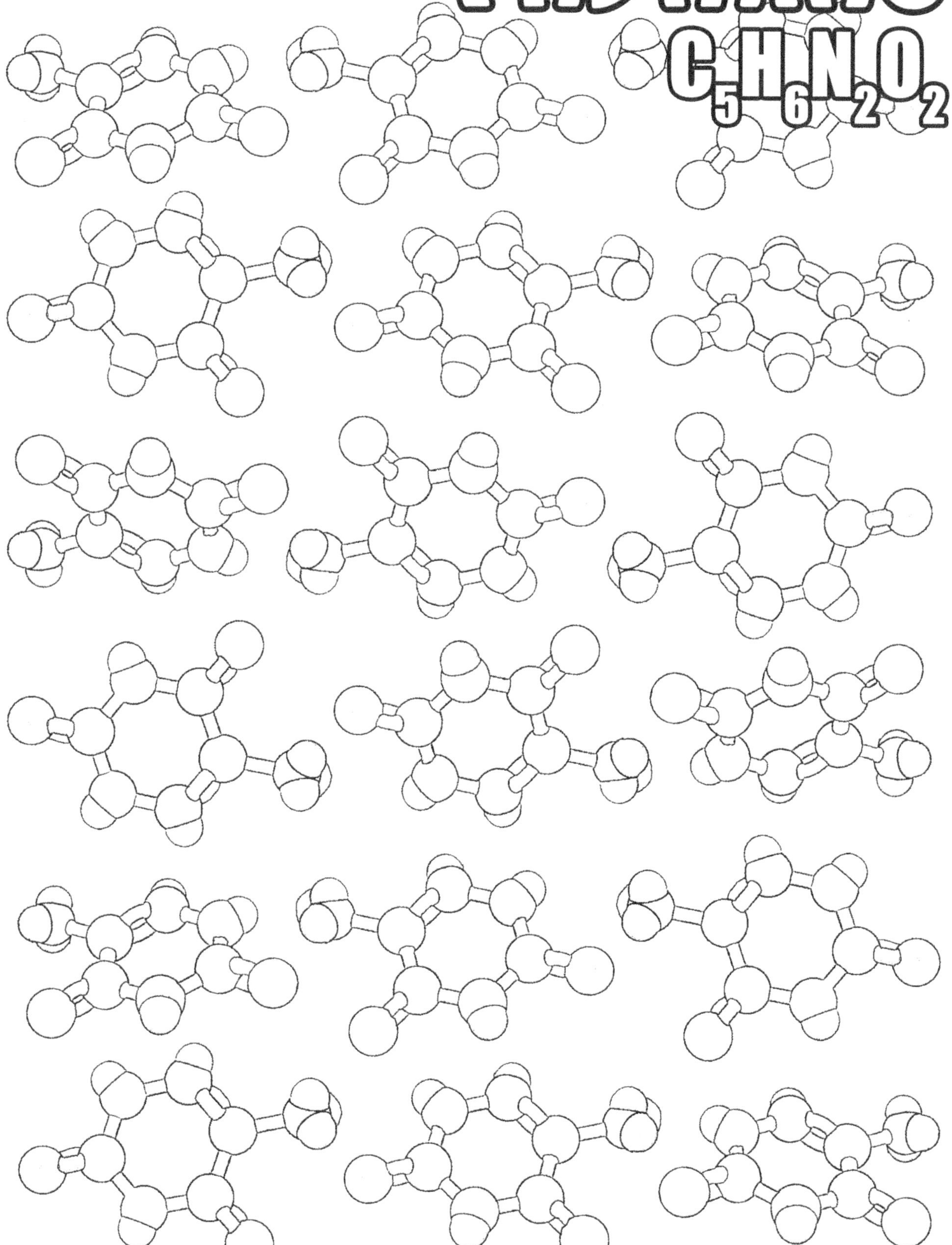

Tocopherol

$C_{29}H_{50}O_2$

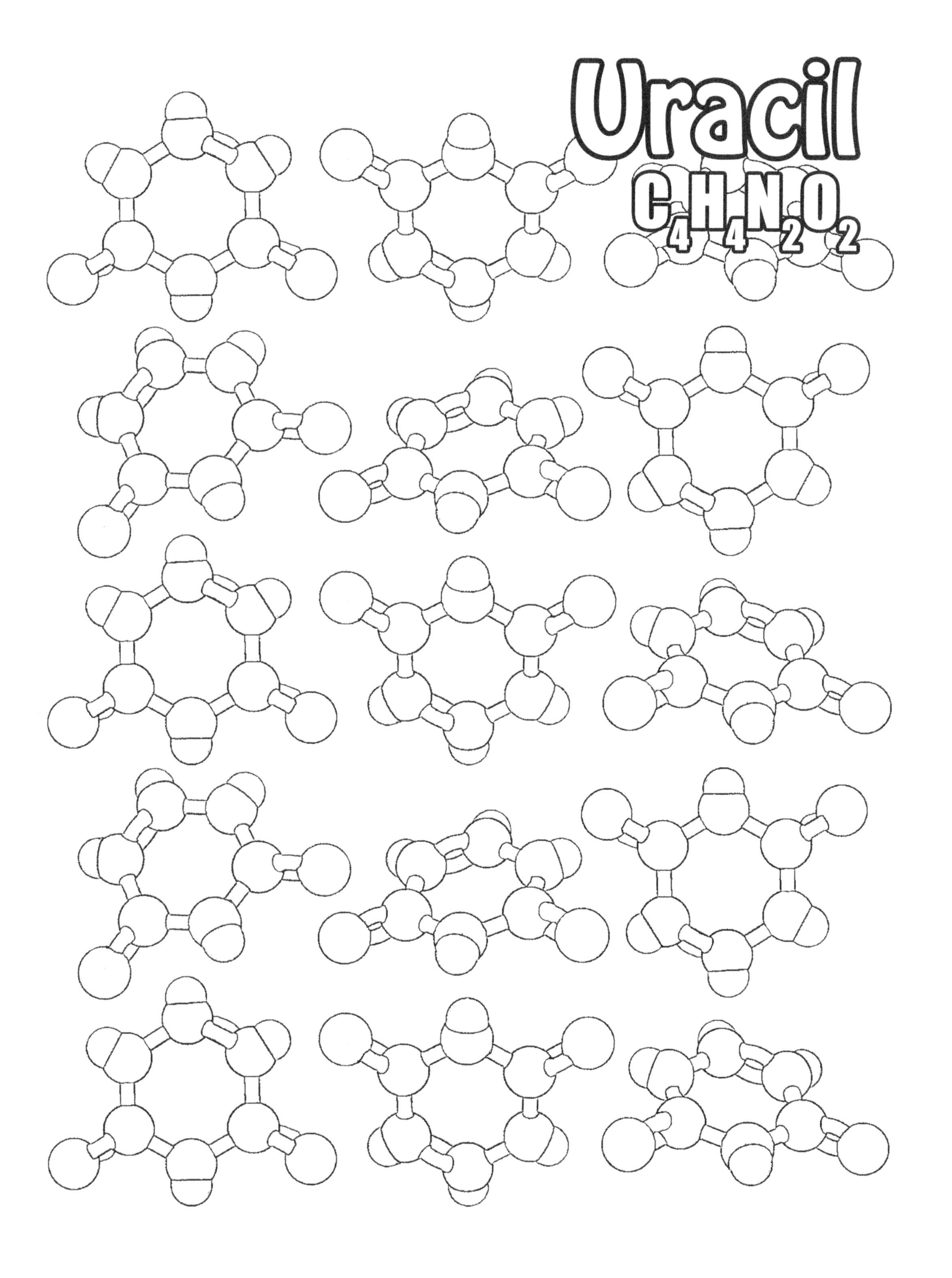
Uracil
$C_4H_4N_2O_2$

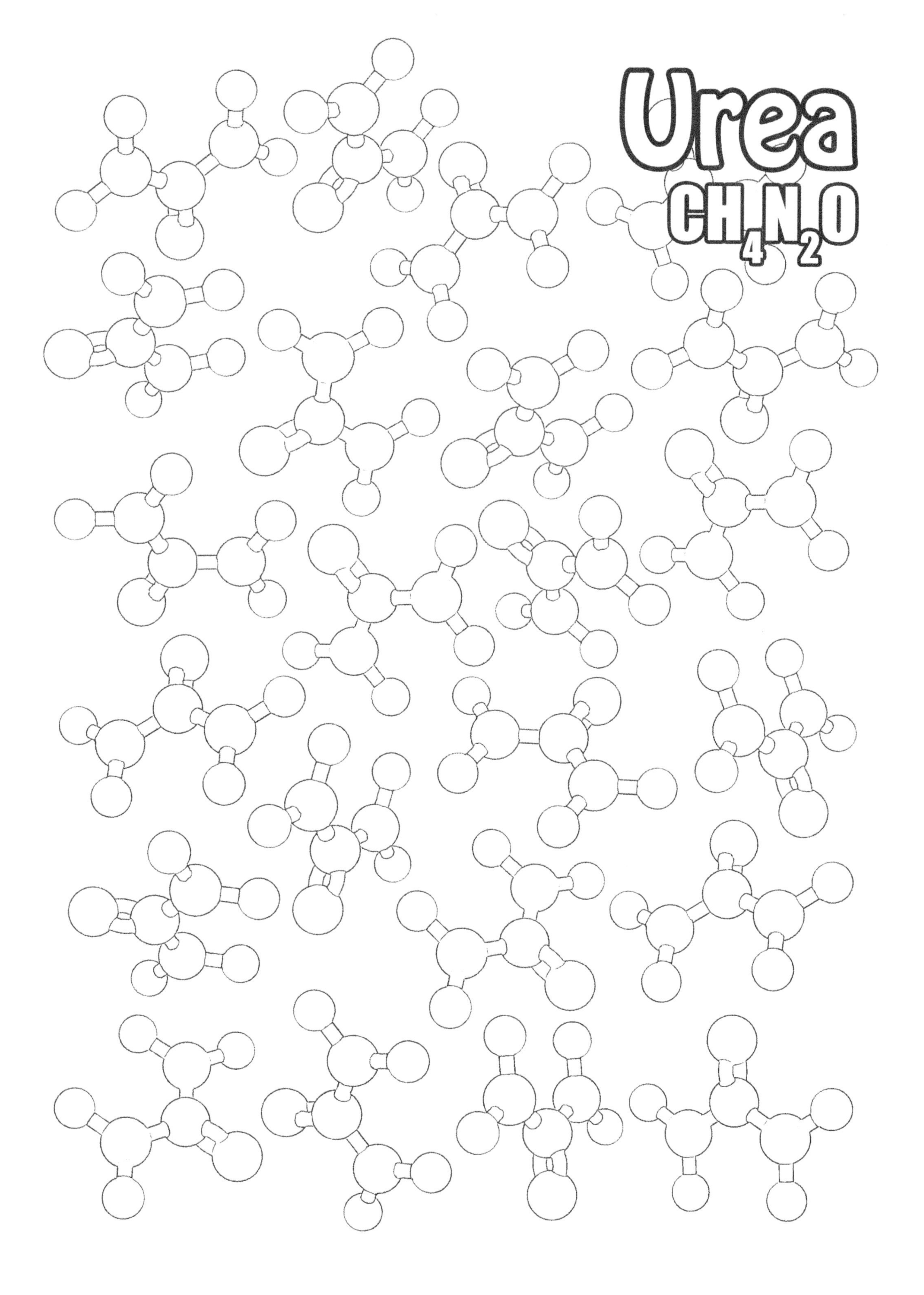
Urea
CH4N2O

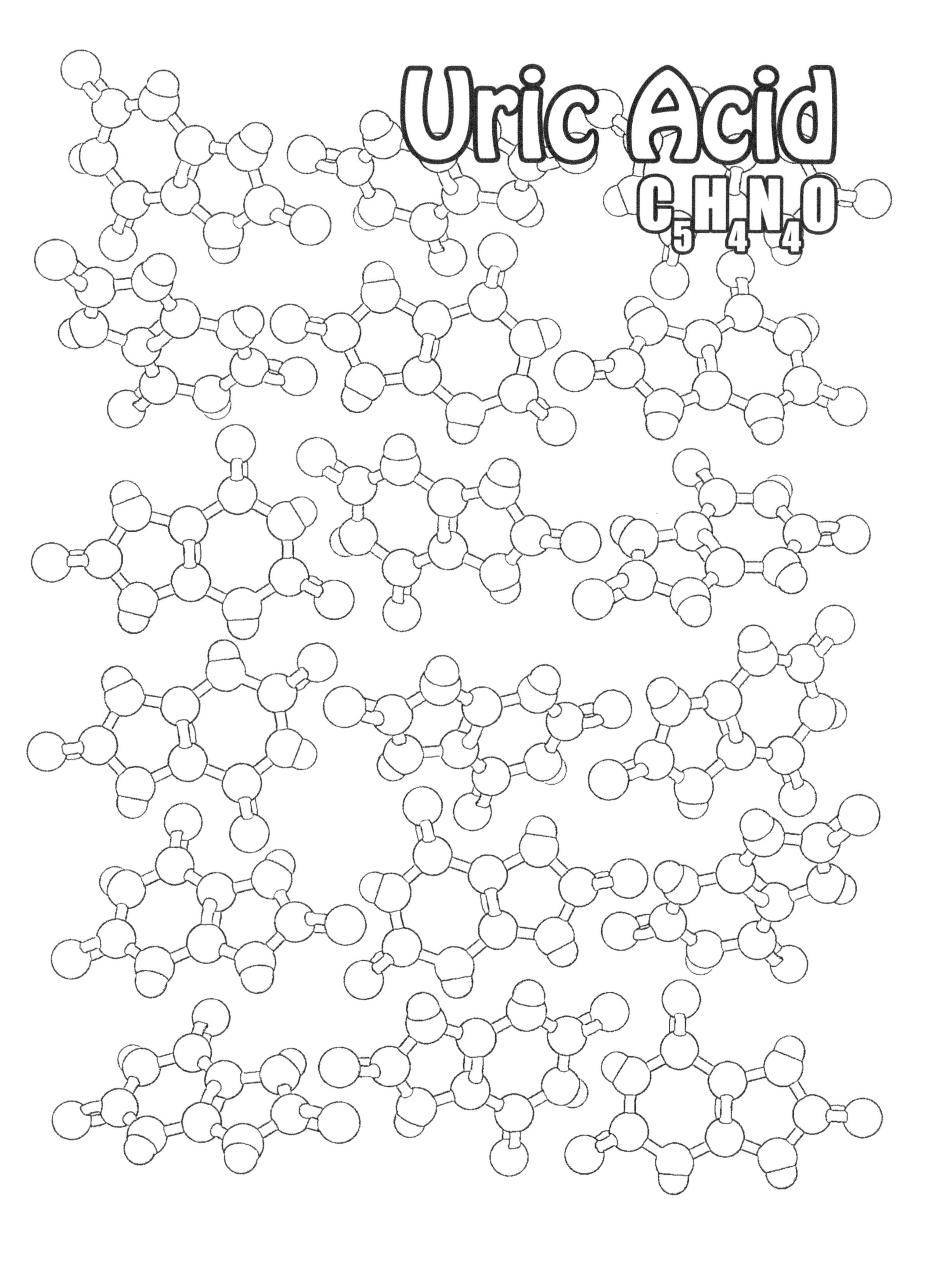
Uric Acid
C5H4N4O

Valerenic Acid

$C_{15}H_{22}O_2$

Valine
$C_5H_{11}NO_2$

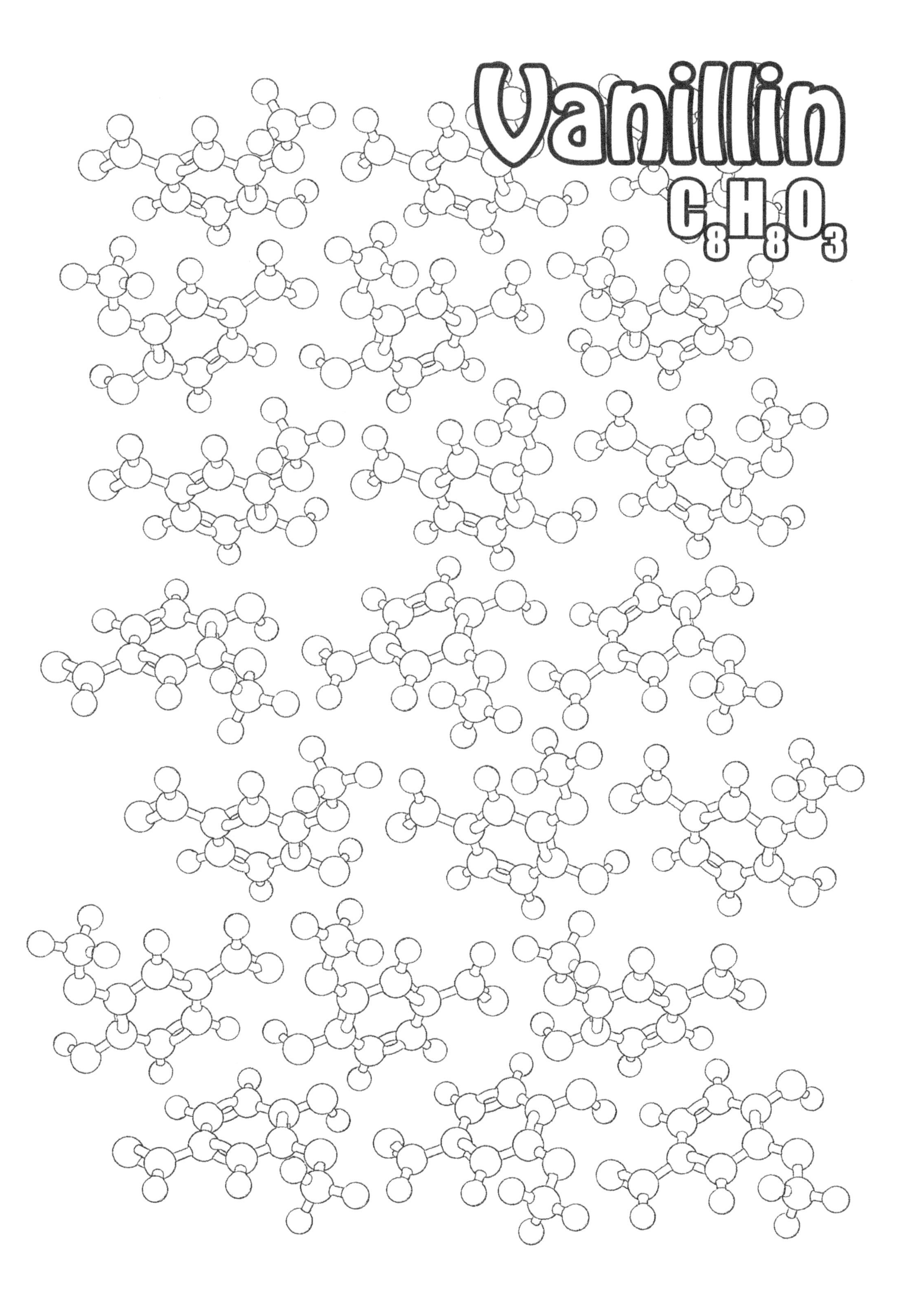
Vanillin
$C_8H_8O_3$

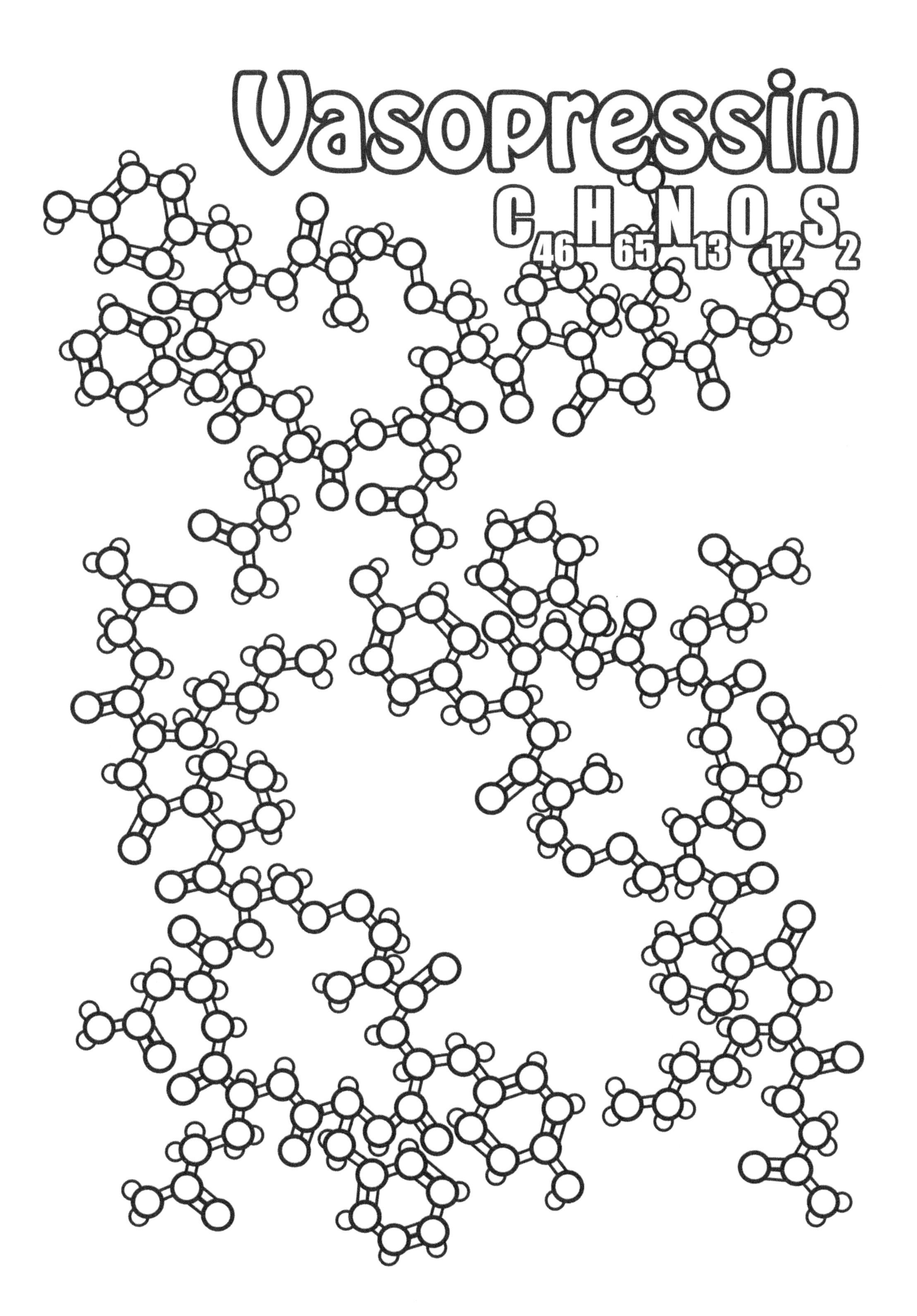

Vasopressin
C46H65N13O12S2

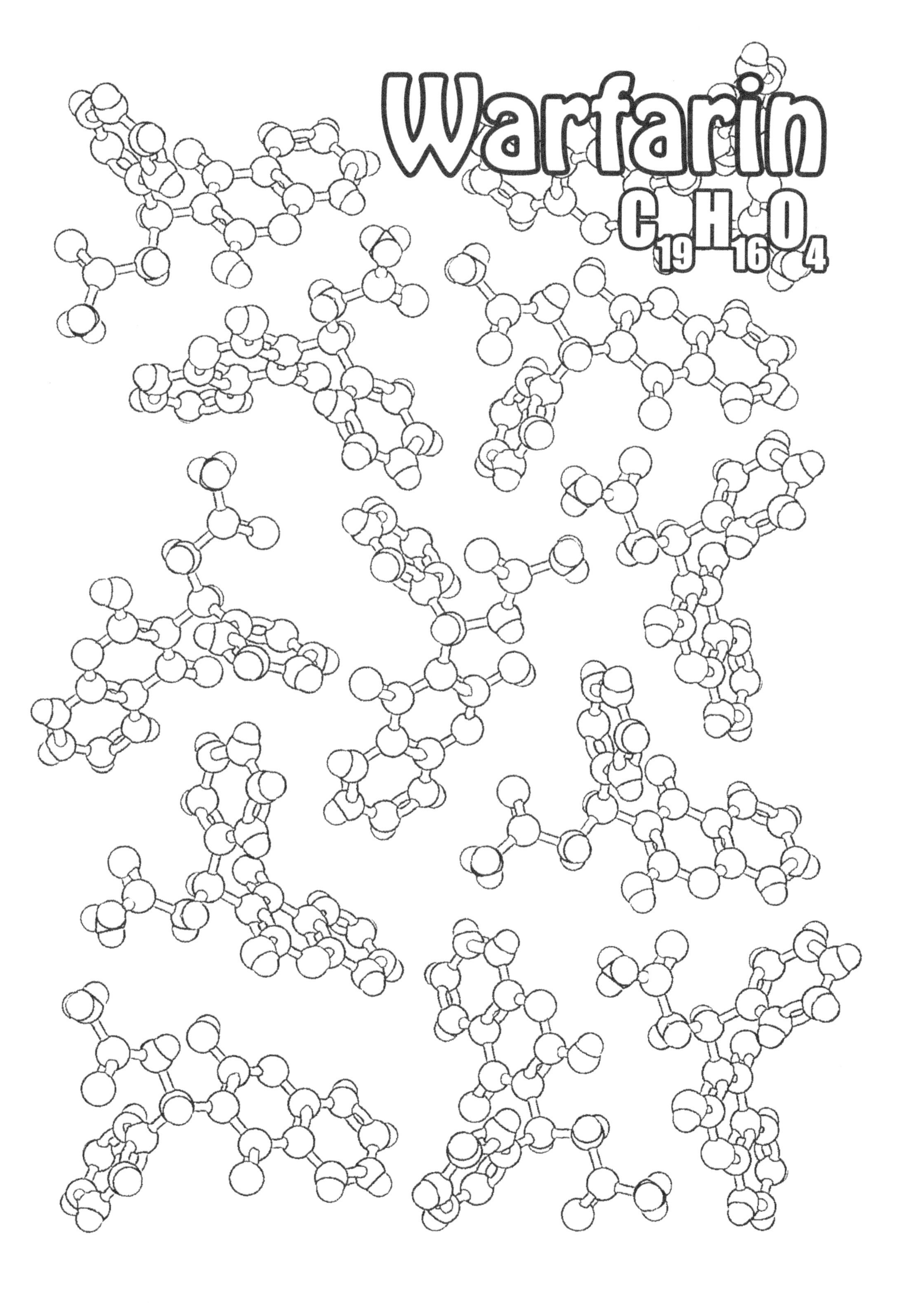
Warfarin
$C_{19}H_{16}O_4$

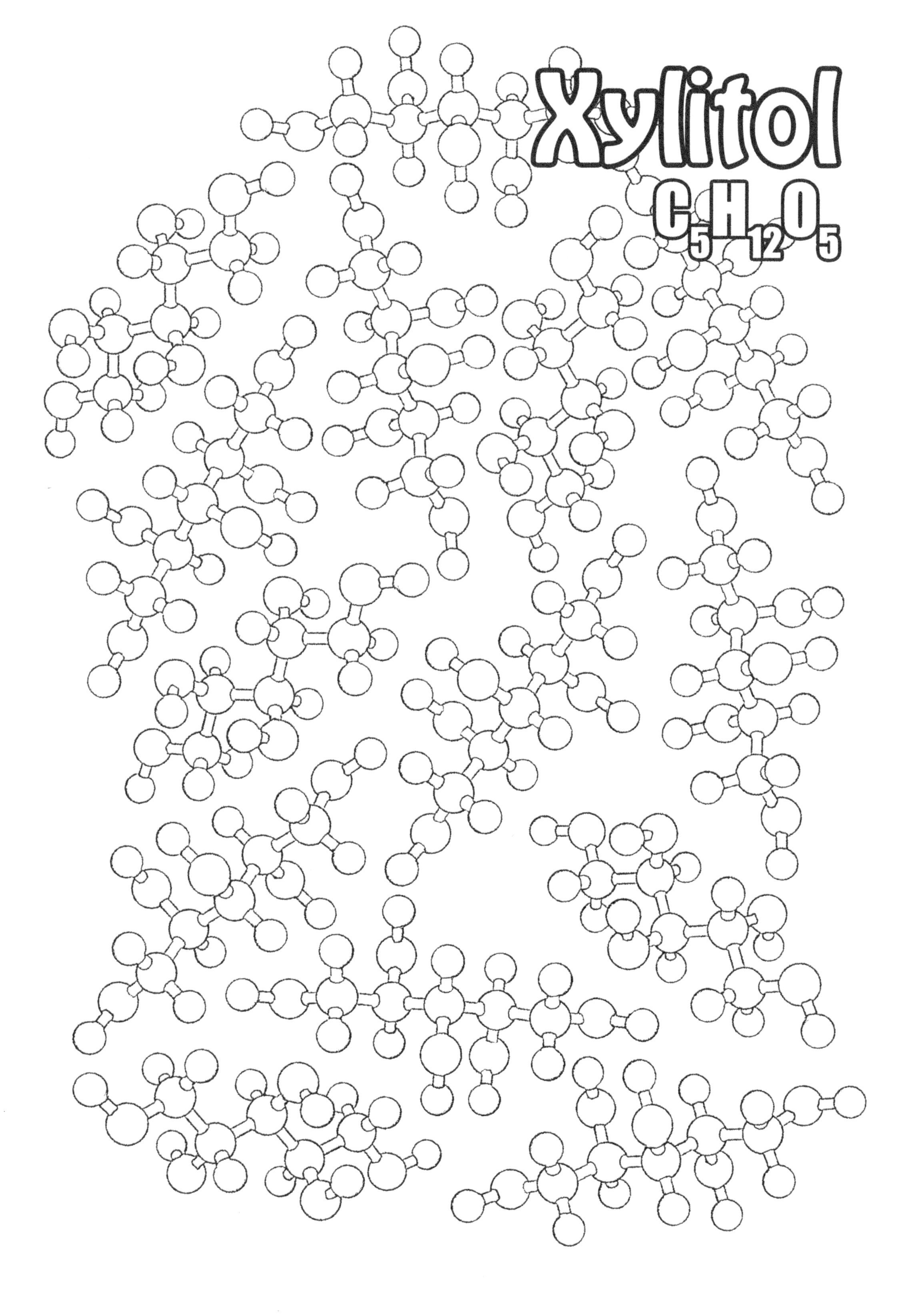
Xylitol
$C_5H_{12}O_5$

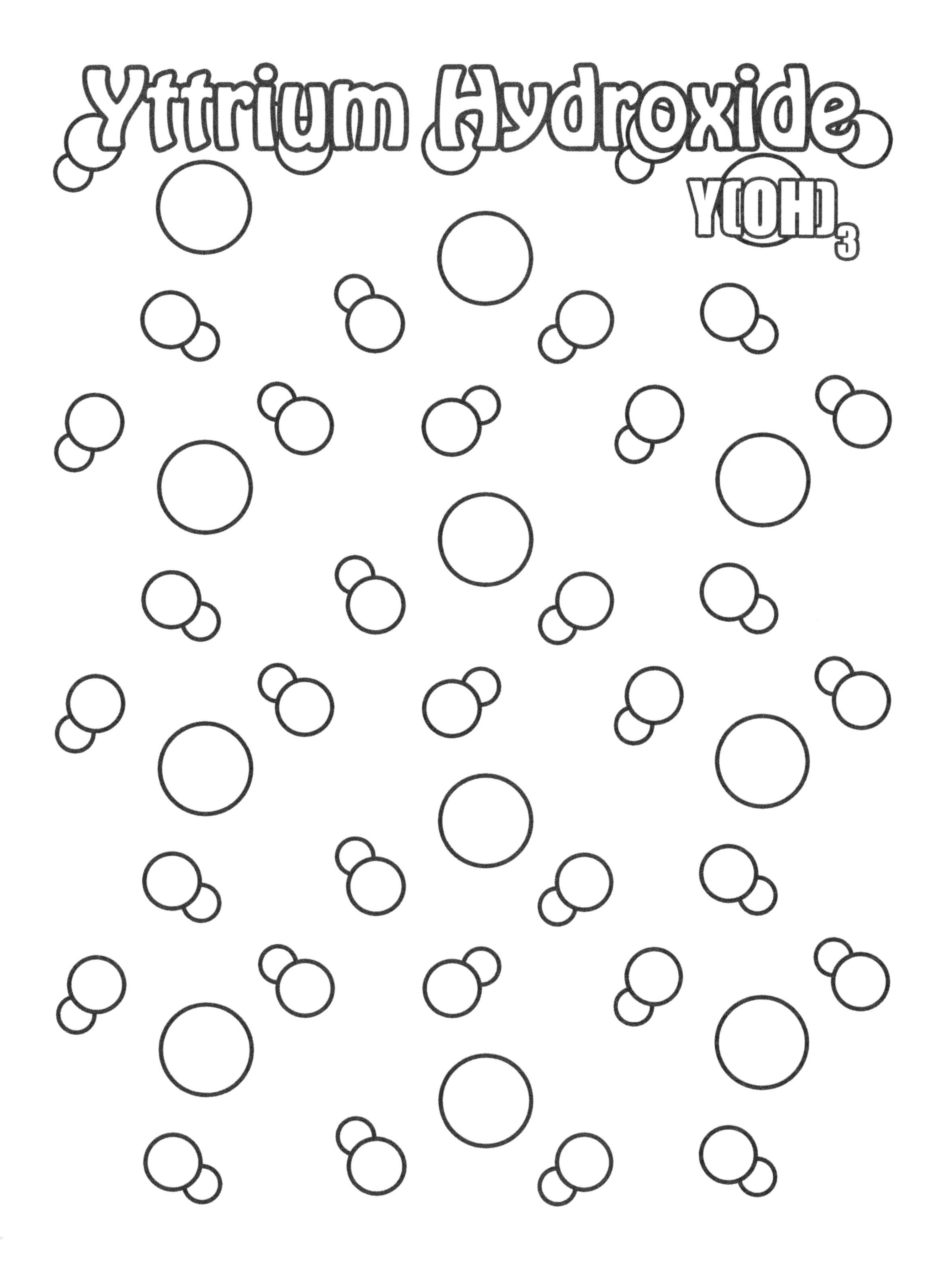
Yttrium Hydroxide
Y(OH)3

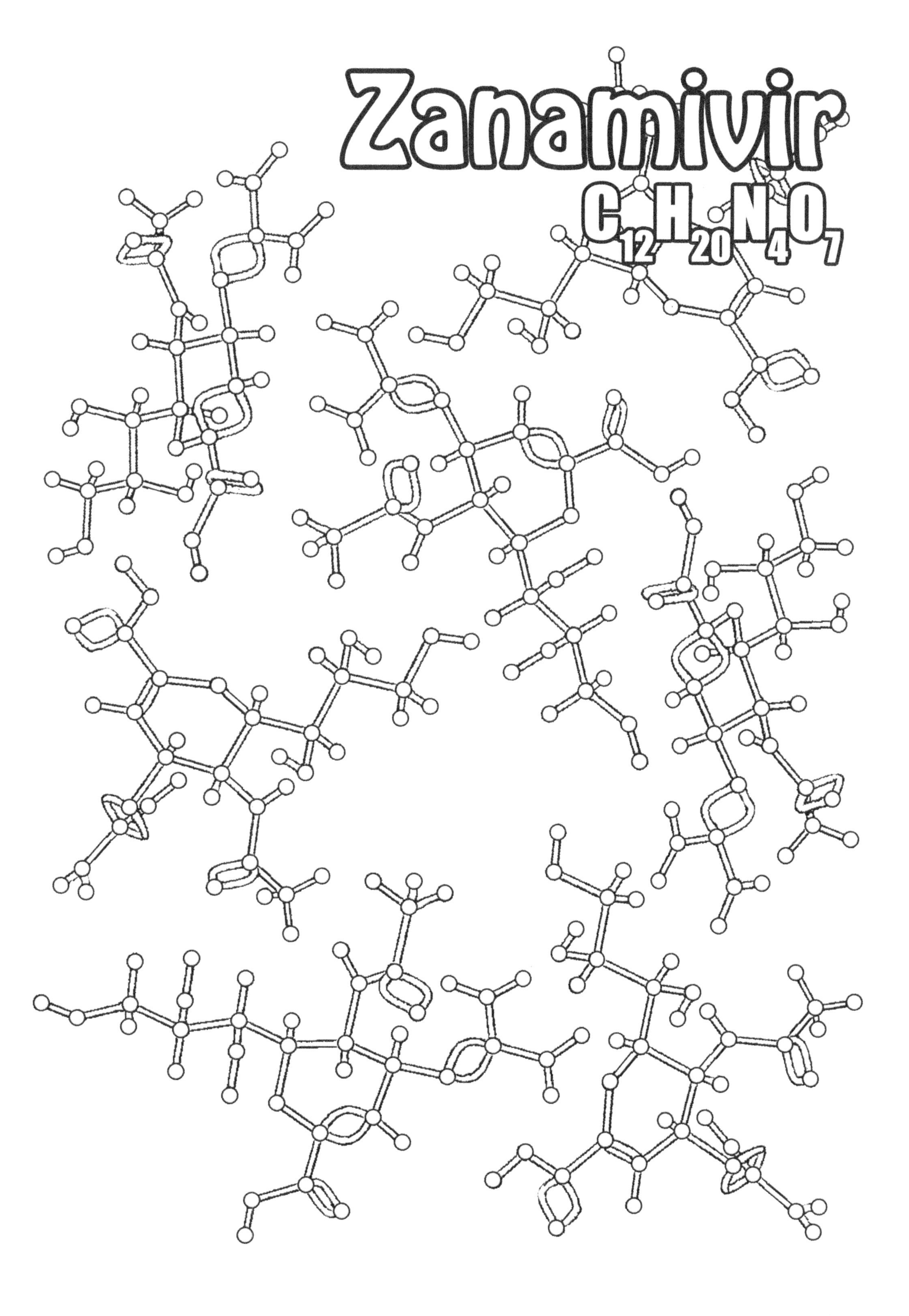
Zanamivir
$C_{12}H_{20}N_4O_7$

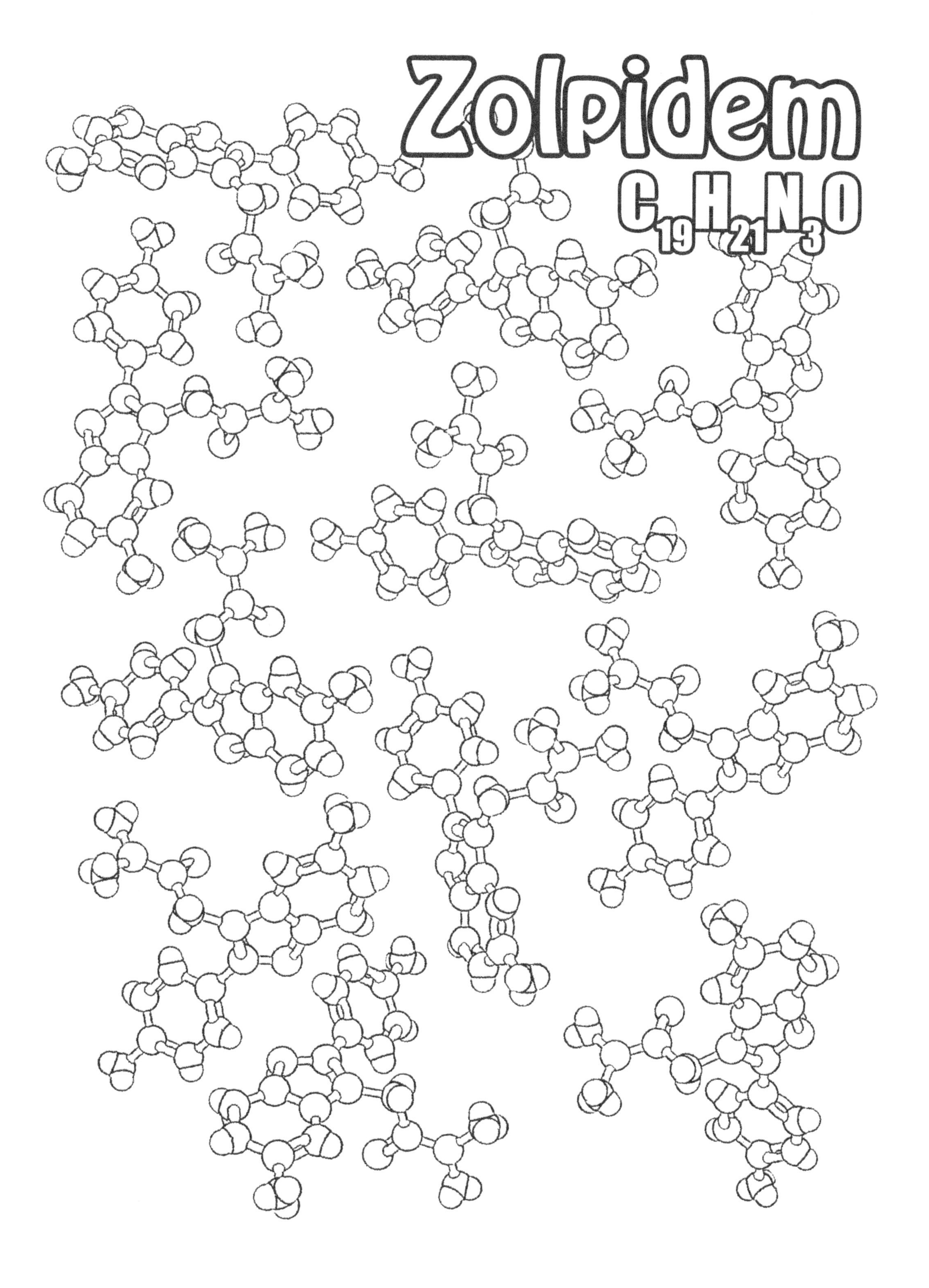
Zolpidem
$C_{19}H_{21}N_3O$

Acetaminophen

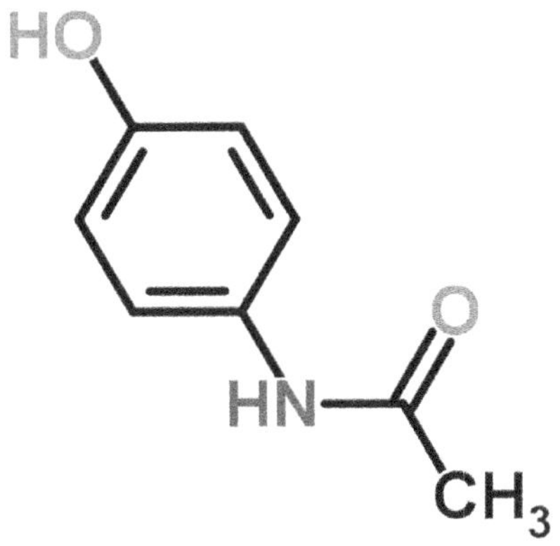

Adenine

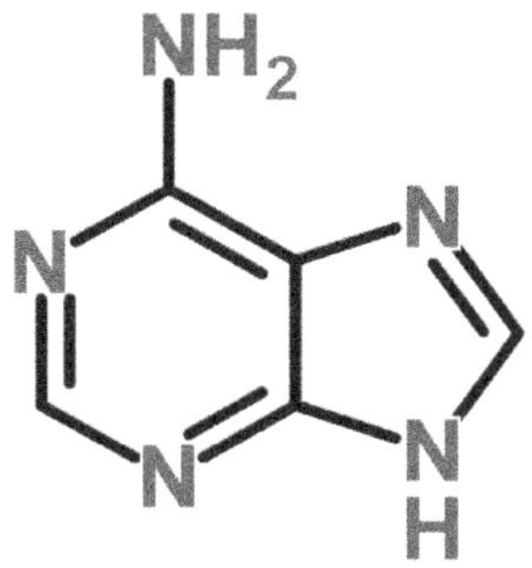

Alprazolam

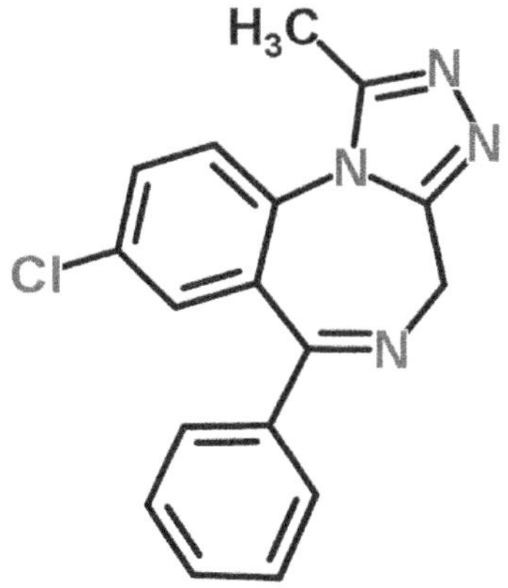

Amoxicillin

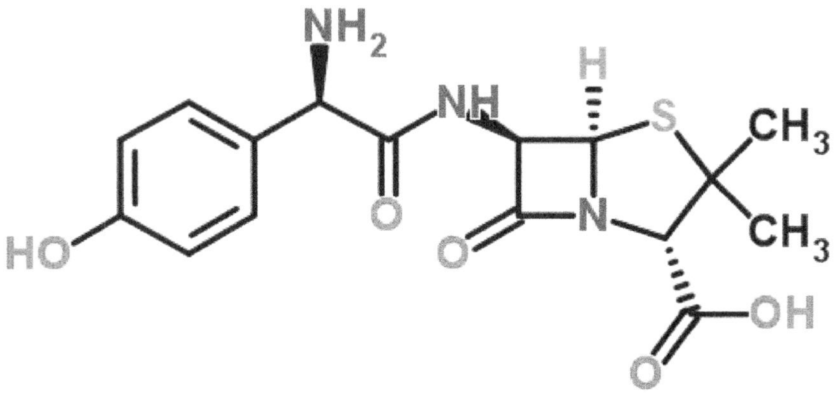

Bacitracin

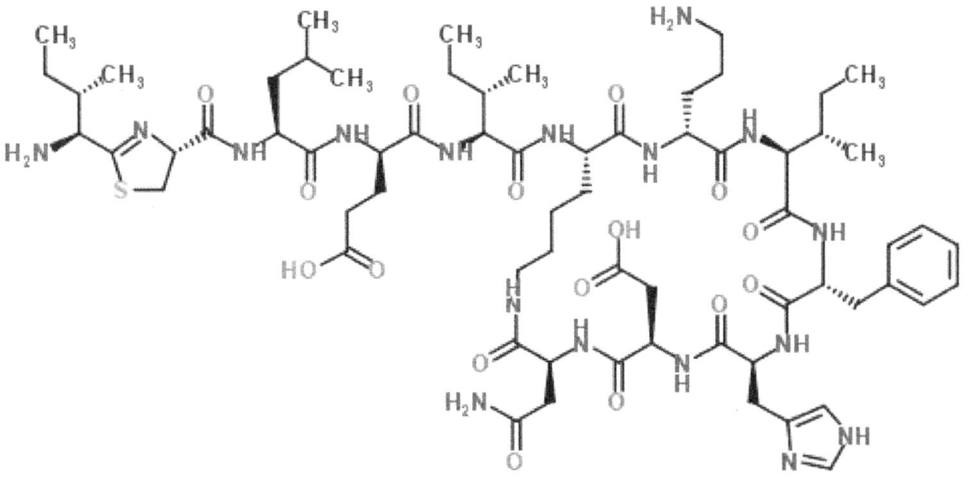

Butadiene

Chloroform

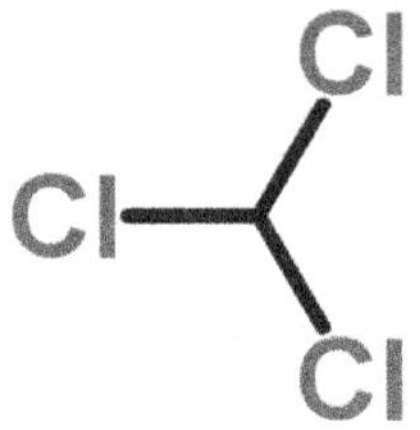

Chlorophyll

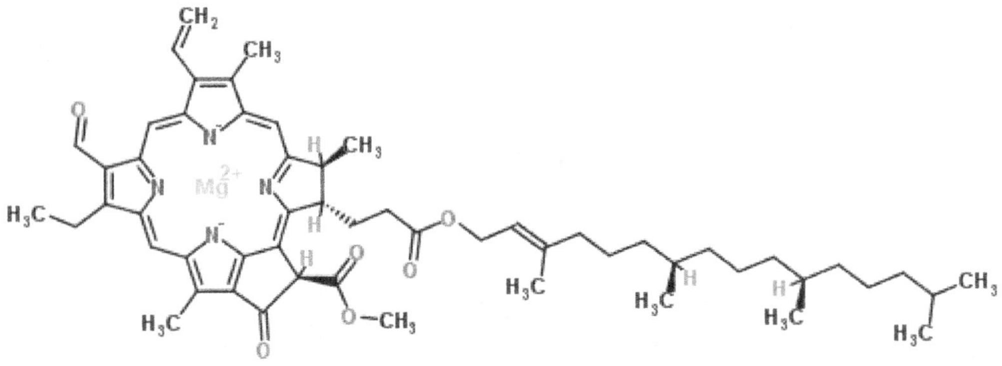

Clonidine

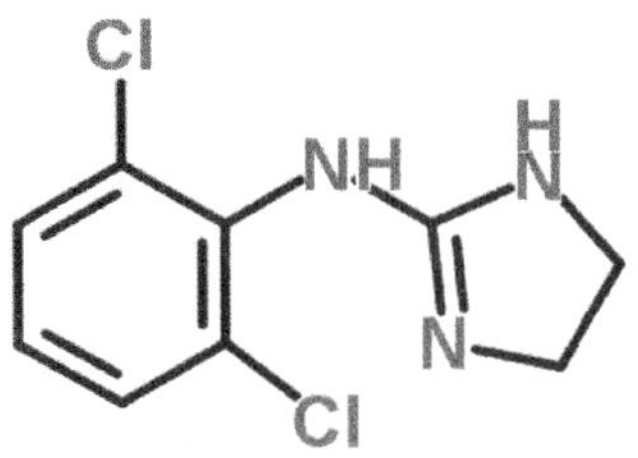

Cytosine

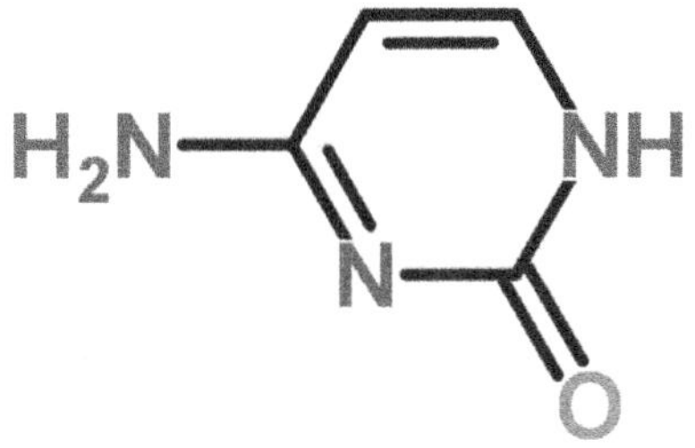

Diazepam

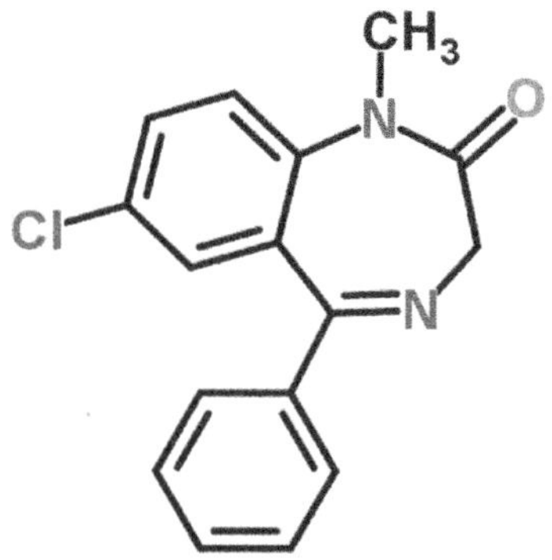

Estazolam

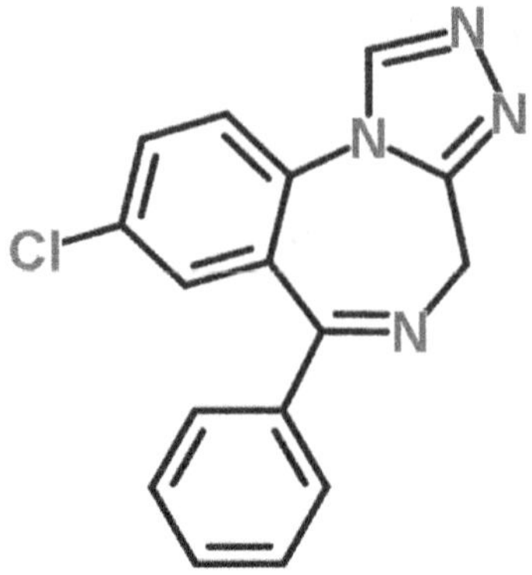

Fluoxetine

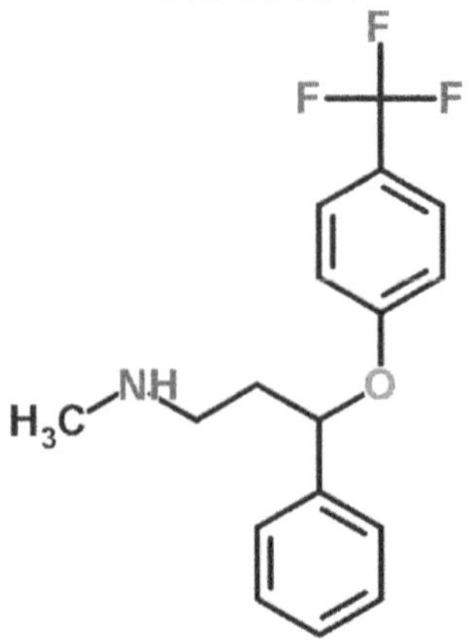

Guanine

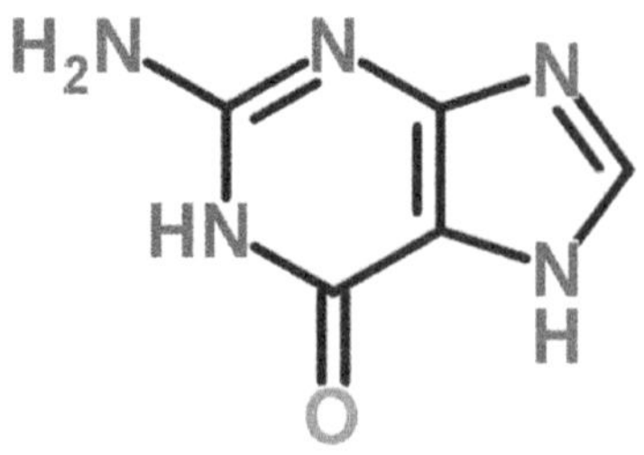

Histamine

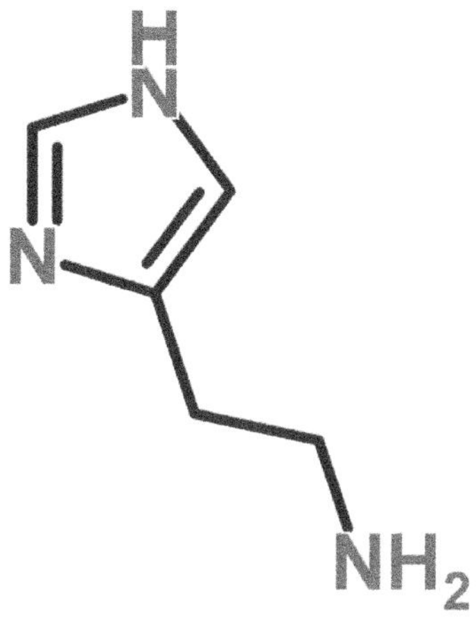

Isoprene

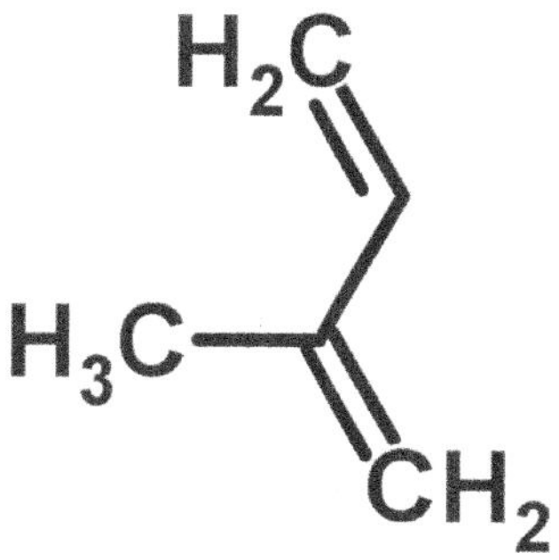

Josamycin

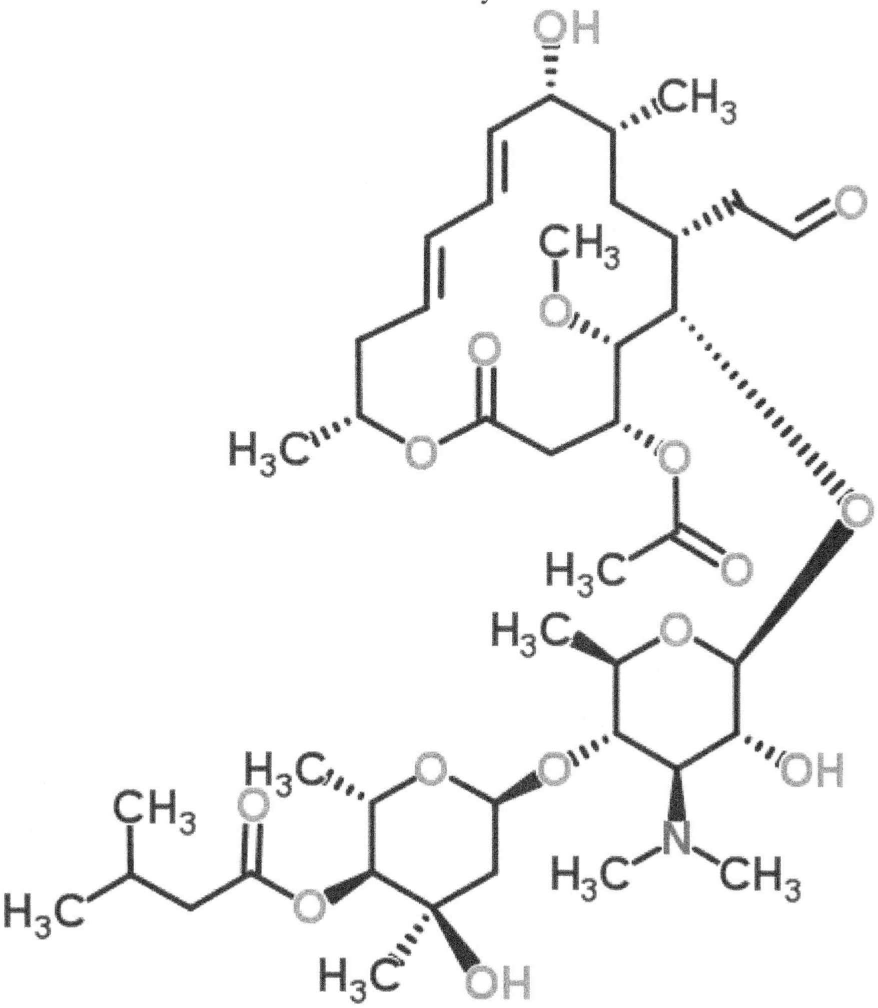

Ketoconazole

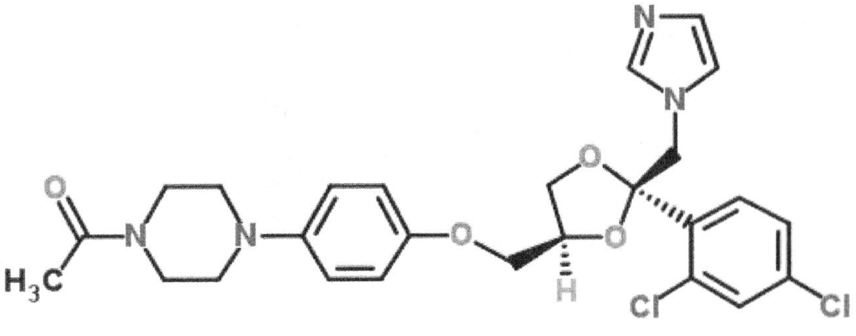

Kojic Acid

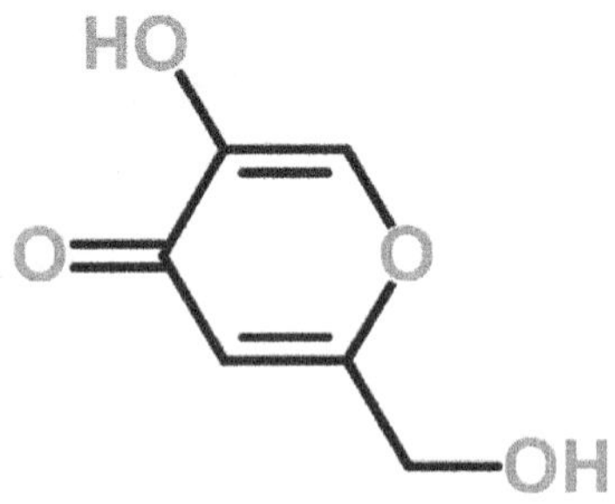

Lorazepam

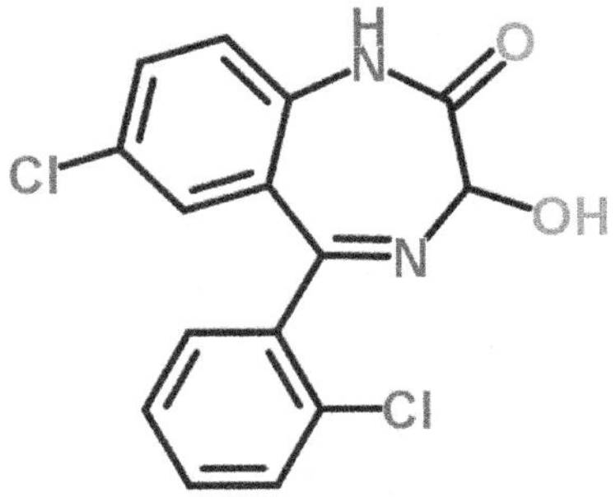

Meclizine

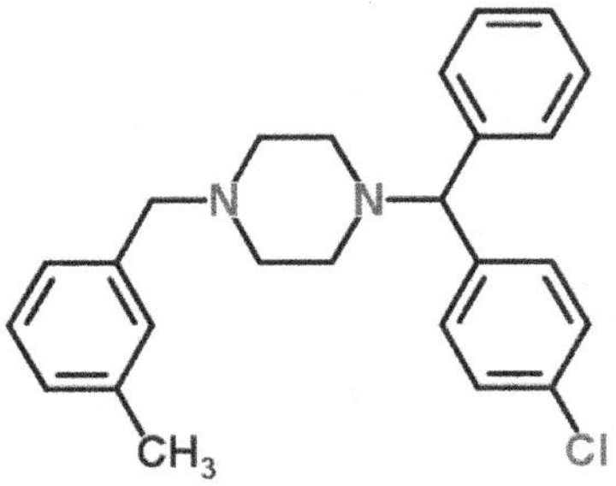

Melanin

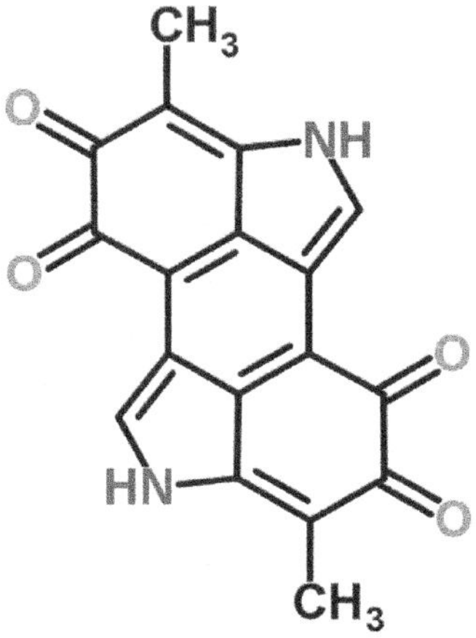

Morphine

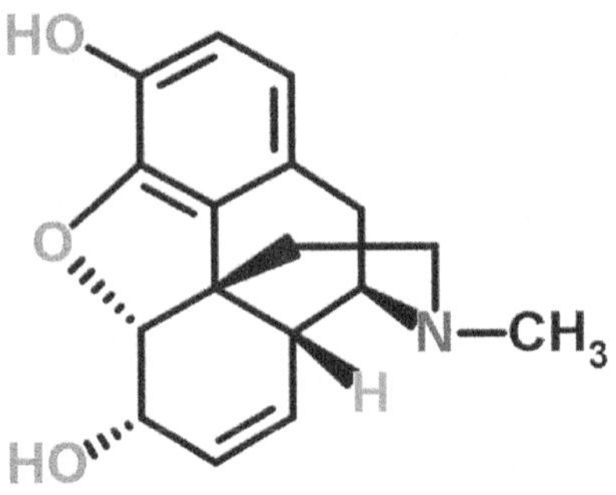

Nicotine

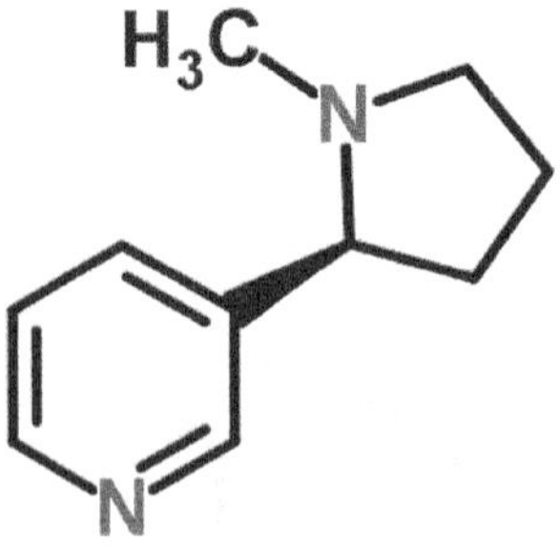

Noradrenaline

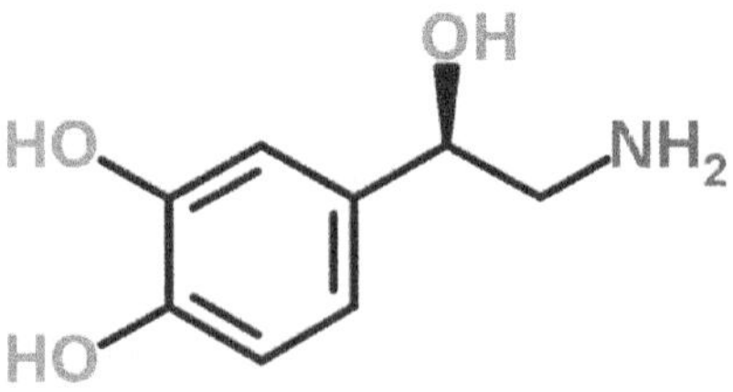

Oxycodone

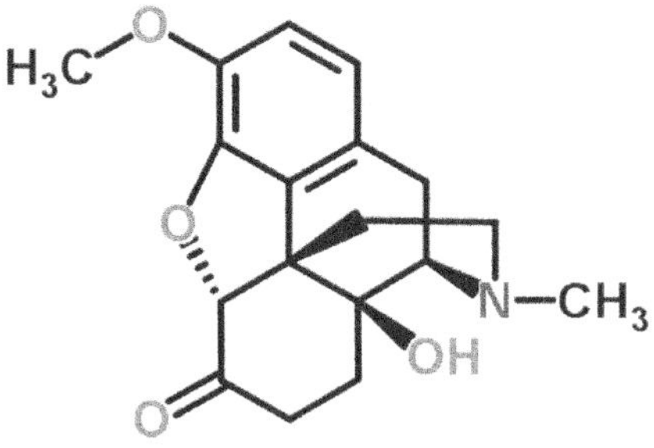

Oxytocin

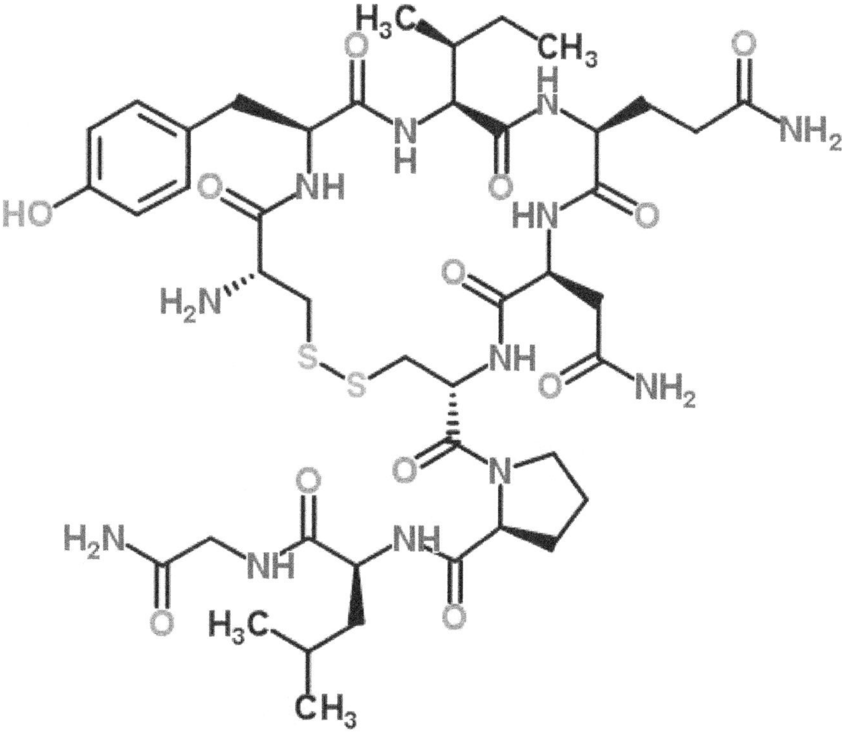

Penicillin

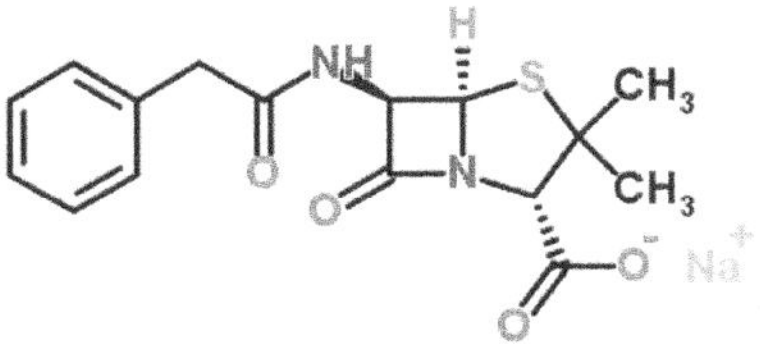

Prazepam

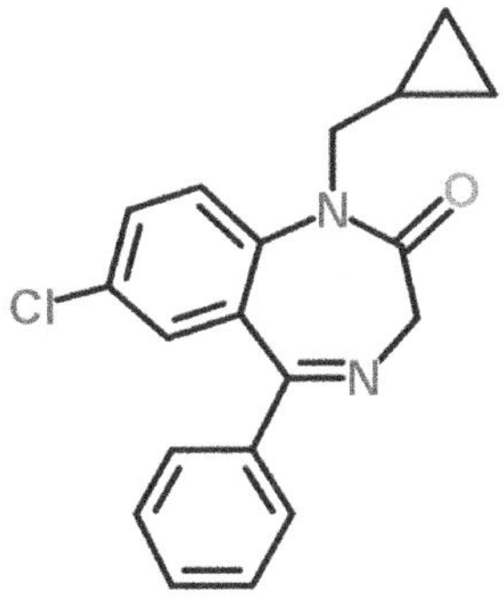

Quazepam

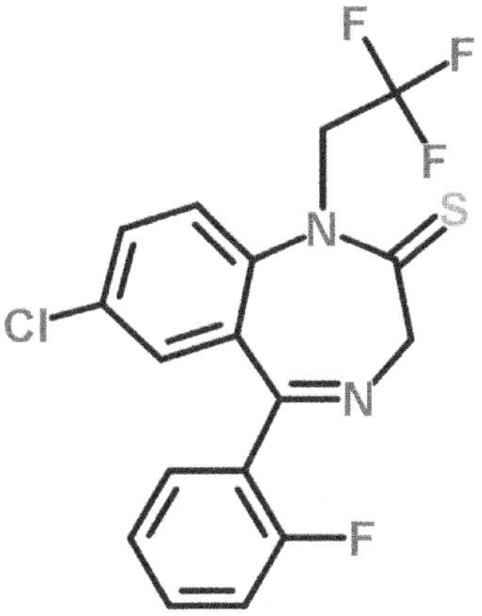

Questin

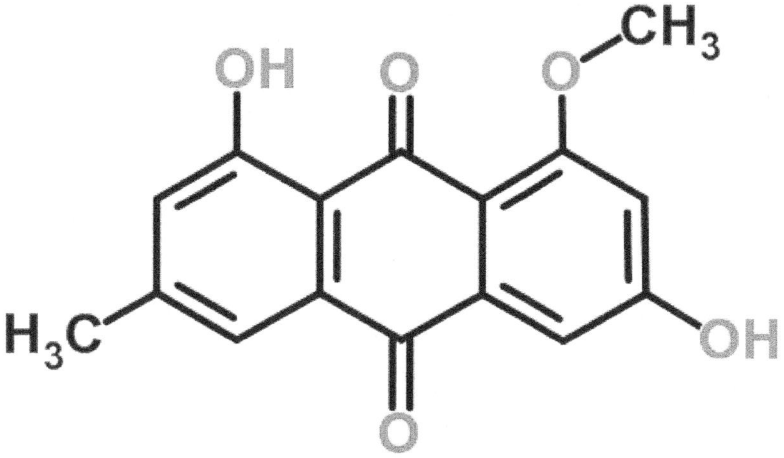

Ranitidine

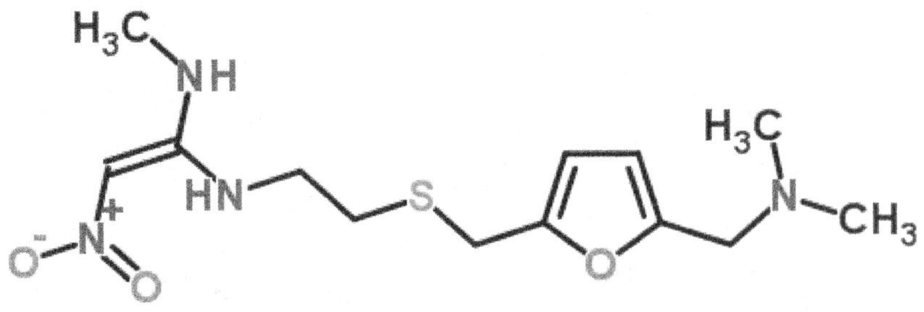

Resveratrol

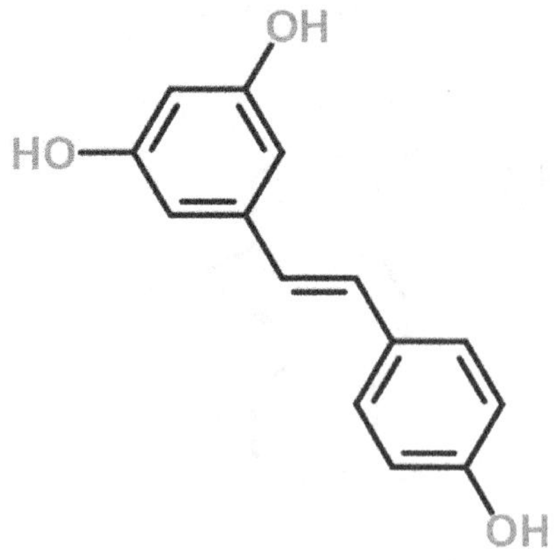

Sertraline

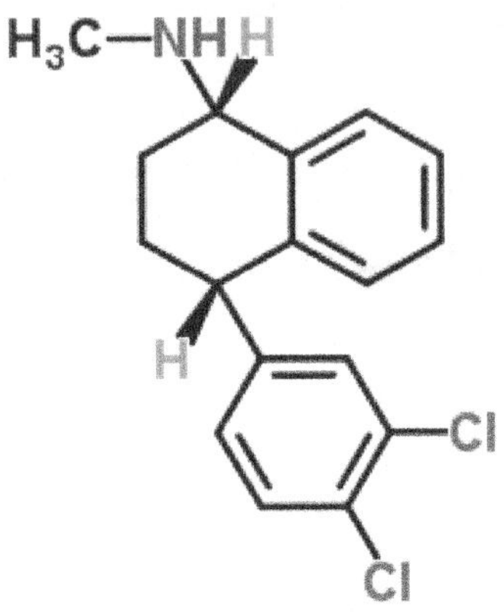

Silymarin

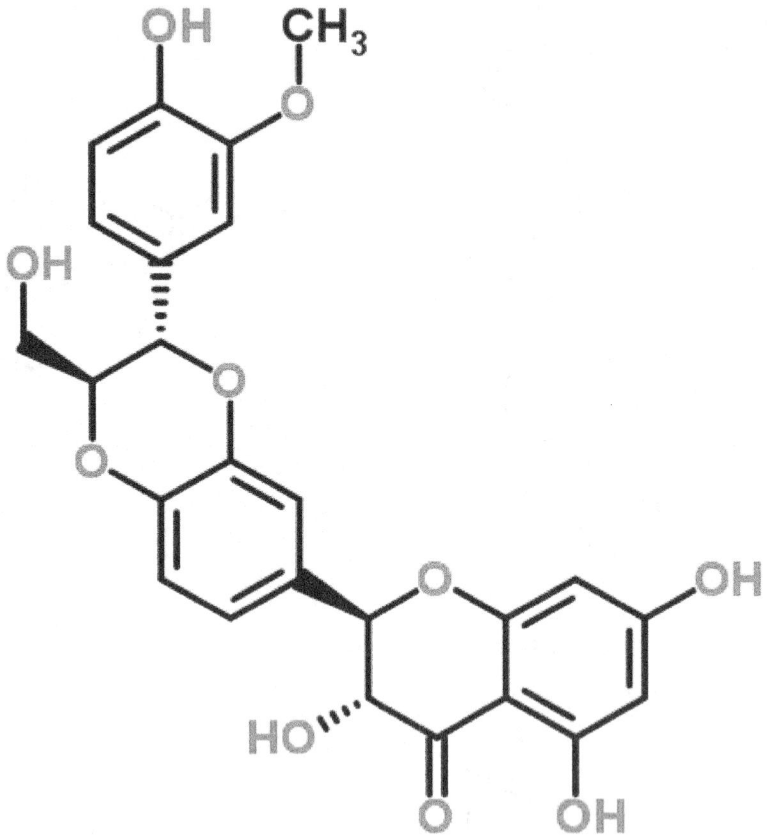

Thiamine

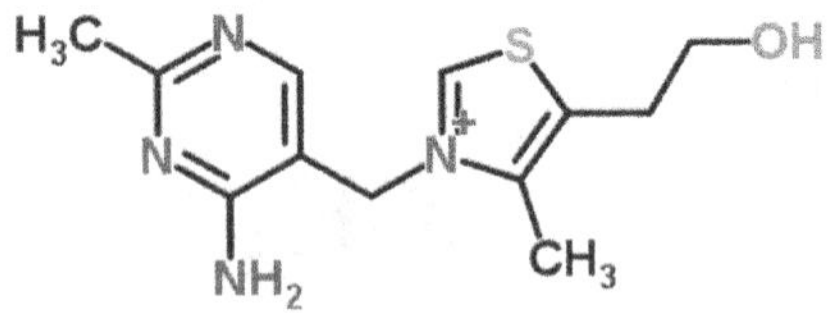

Thymine

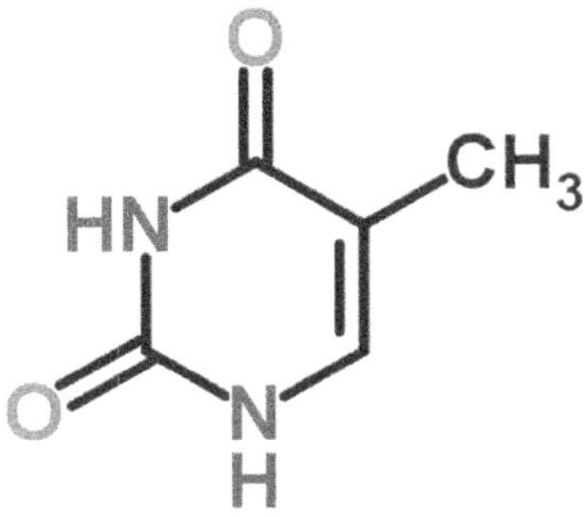

Uracil

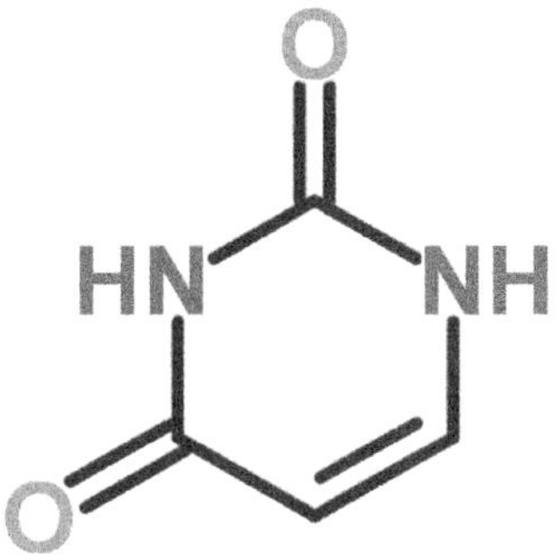

Urea

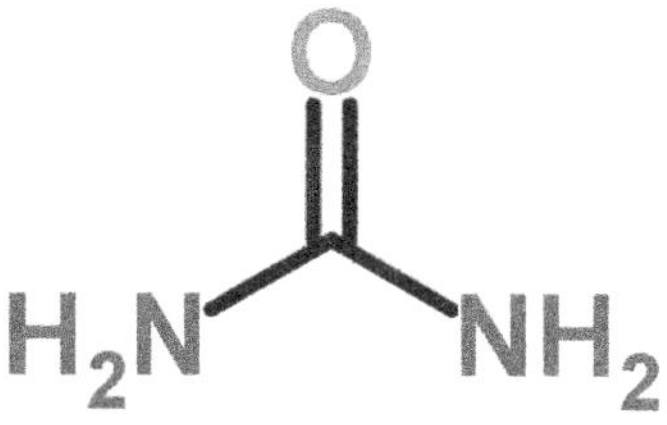

Uric Acid

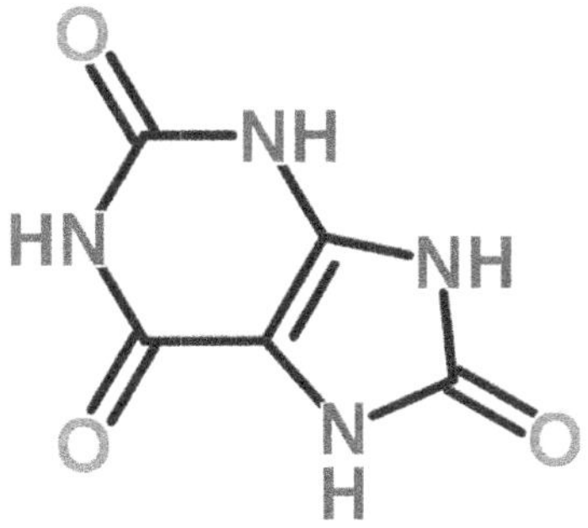

Valerenic Acid

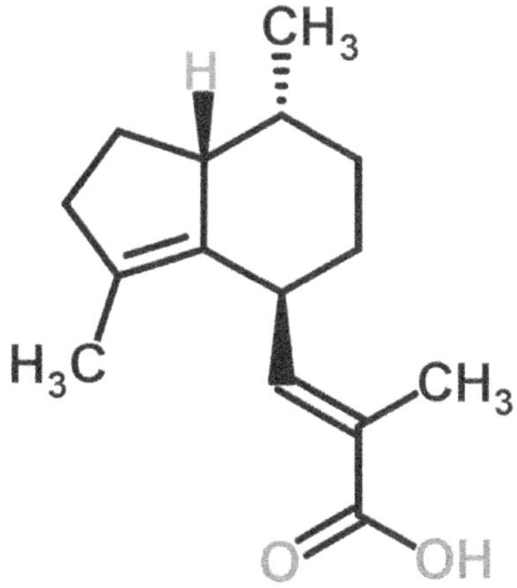

Valine

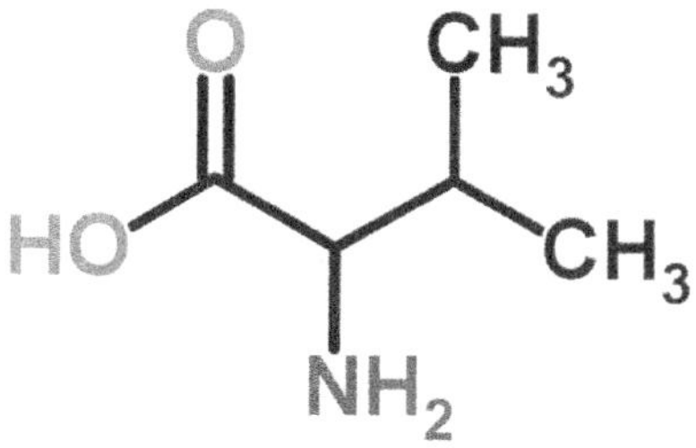

Vanillin

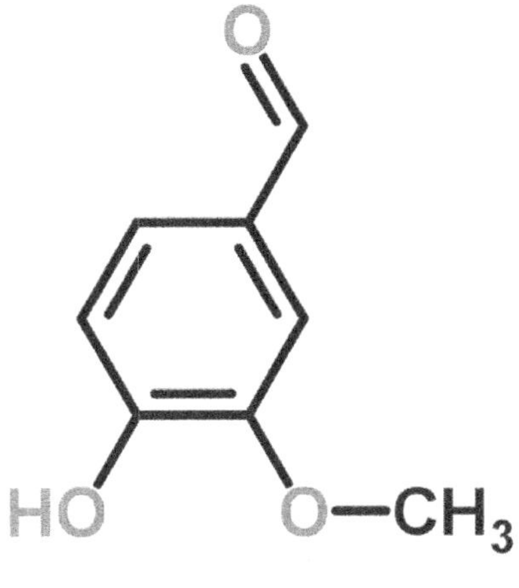

Vasopressin

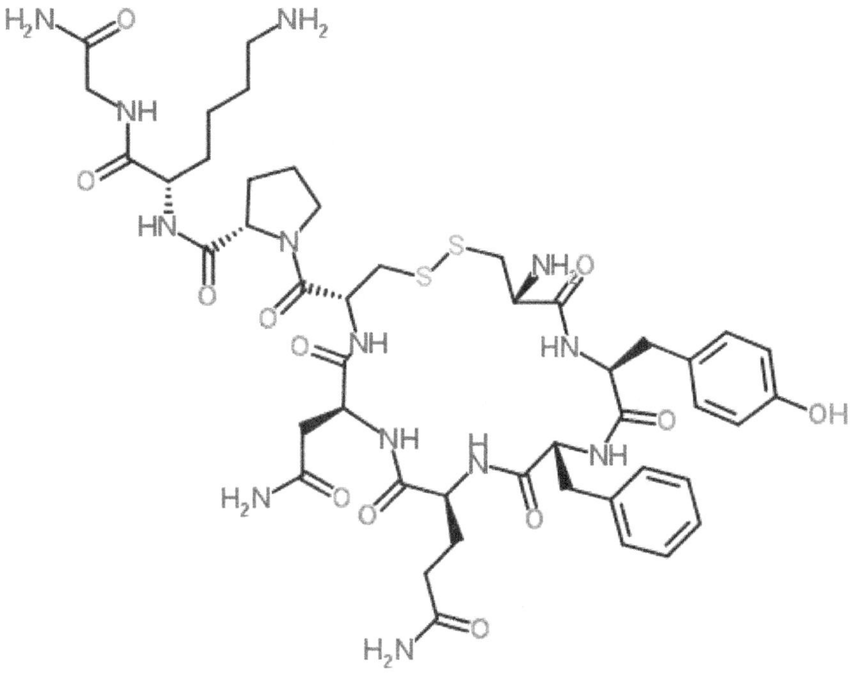

Warfarin

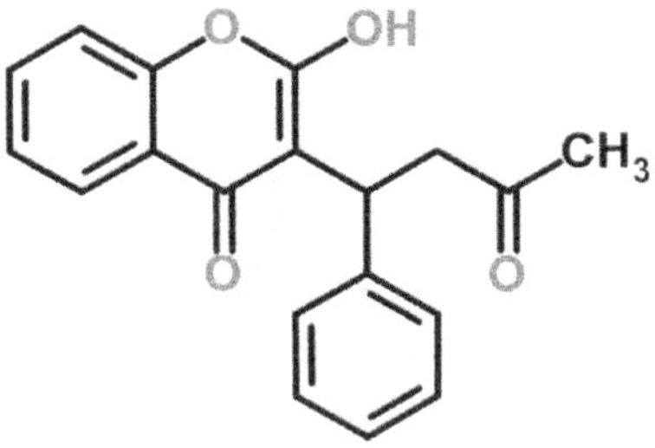

Xylitol

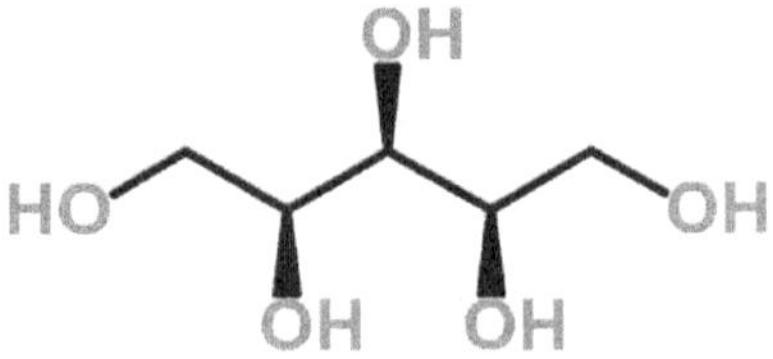

Yttrium Hydroxide

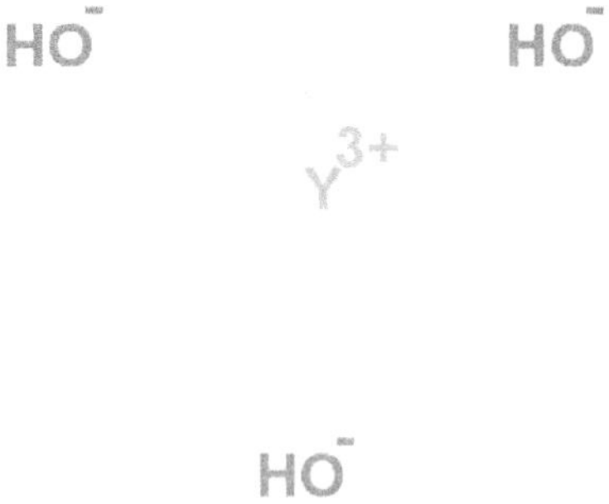

Zanamivir

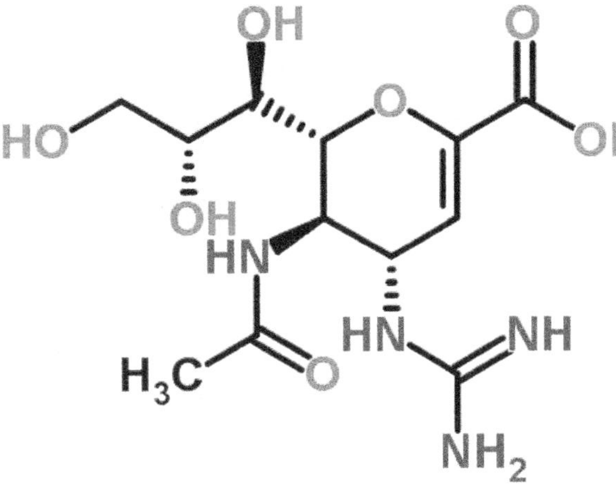

Zolpidem

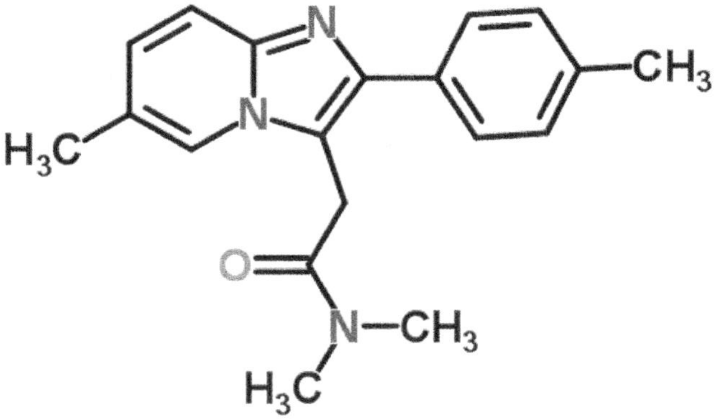

Skeletal formula images are used with permission from ChemSpider.com

Coloring pages were created showing structures from various angles. In more complex structures, some atoms and bonds might only be visible from certain angles and not visible from others.

Certain aspects of chemistry such as resonance, ionic bonds, and stereochemistry can be difficult to accurately convey in coloring pages. While attempts were made to make this book as accurate as possible, it is recommended that those who want a more complete understanding of these structures use other resources in addition to this book, as it is beyond the scope of this book to give a complete portrayal of the complexities these structures represent.

Made in the USA
Monee, IL
19 November 2024